Brain Health for Learning

Brain Health for Learning

What neuroscience brings to education?

Denis Staunton and Aimie Brennan

PETER LANG
Oxford · Berlin · Bruxelles · Chennai · Lausanne · New York

Bibliographic information published by the Deutsche Nationalbibliothek. The German National Library lists this publication in the German National Bibliography; detailed bibliographic data is available on the Internet at http://dnb.d-nb.de.

A catalogue record for this book is available from the British Library.

Library of Congress Cataloging-in-Publication Data

Names: Staunton, Denis, author. | Brennan, Aimie, author.
Title: Brain Health for learning: what neuroscience brings to
    education? / Denis Staunton, Aimie Brennan.
Description: Oxford, United Kingdom; New York, NY: Peter Lang, 2024. |
    Includes bibliographical references and index.
Identifiers: LCCN 2024028818 (print) | LCCN 2024028819 (ebook) | ISBN
    9781803741055 (paperback) | ISBN 9781803741062 (ebook) | ISBN
    9781803741079 (epub)
Subjects: LCSH: Cognitive neuroscience. | Brain.
Classification: LCC QP360.5 .S73 2024 (print) | LCC QP360.5 (ebook) | DDC
    612.8/233--dc23/eng/20240705
LC record available at https://lccn.loc.gov/2024028818
LC ebook record available at https://lccn.loc.gov/2024028819

All illustrations and figures have been licensed through purchase from Adobe Stock Images.

Cover image by NE97/stock.adobe.com
Cover design by Peter Lang Group AG

ISBN 978-1-80374-105-5 (print)
ISBN 978-1-80374-106-2 (ePDF)
ISBN 978-1-80374-107-9 (ePub)
DOI 10.3726/b20604

© 2024 Peter Lang Group AG, Lausanne
Published by Peter Lang Ltd, Oxford, United Kingdom
info@peterlang.com – www.peterlang.com

Denis Staunton and Aimie Brennan have asserted their right under the Copyright, Designs and Patents Act, 1988, to be identified as Authors of this Work.

All rights reserved.
All parts of this publication are protected by copyright.
Any utilisation outside the strict limits of the copyright law, without the permission of the publisher, is forbidden and liable to prosecution.
This applies in particular to reproductions, translations, microfilming, and storage and processing in electronic retrieval systems.

This publication has been peer reviewed.

I dedicate this book to my ever-supporting wife Sheelagh, and our four children; Ciara, Ronan, Aideen and Lyndsey. To my extended family Steven, Ian and Grainne, and my beloved and cherished grandchildren, Sophie, Noah, Julia, Lucas, Amelie, Andrew, Robert and Alice. Many thanks for contributing to my brain health.

Denis

I dedicate this book to the memory of my grandfather who always believed I would write a book but unfortunately didn't get the chance to read it. His light burned brightly.

Aimie

Contents

List of Figures	xi
Preface	xiii
Acknowledgements	xv
Introduction	1
SECTION 1	9
CHAPTER 1 Meet Your Brain	11
CHAPTER 2 The Architecture of the Brain	23
CHAPTER 3 The Importance of Neurons	41
CHAPTER 4 The Power of Plasticity	55
SECTION 2	69
CHAPTER 5 What Is Educational Neuroscience?	71

CHAPTER 6 How Learning Happens	85
CHAPTER 7 Harnessing Habits for Learning	113
SECTION 3	127
CHAPTER 8 Learning across the Life Cycle	129
CHAPTER 9 Interpersonal Relationships	147
CHAPTER 10 Food for Thought	165
CHAPTER 11 Exercise and Neurogenesis	181
CHAPTER 12 Sleep and Brain Health	197
CHAPTER 13 Why Targets Focus the Brain	213
CHAPTER 14 Yearning for Meaning	227
CHAPTER 15 Liquids: Elixirs of Life	241

CHAPTER 16
Empathy for Brain Health   255

Glossary   267

Index   281

Figures

Figure 1:	A Phronelogical Map	18
Figure 2:	Human Brain Top View	25
Figure 3:	Human Brain Four Lobes	28
Figure 4:	Human Brain: Saggital Section (Interior)	31
Figure 5:	Human Brain: The Limbic System	34
Figure 6:	The Central Nervous System	42
Figure 7:	The Structure of a Neuron	44
Figure 8:	A Neural Network	49
Figure 9:	Neuroplasticity: Synaptic Density	57
Figure 10:	Standard vs Enriched Environment	59
Figure 11:	The Learning Zone	64
Figure 12:	The Working Memory	90
Figure 13:	The Habit Loop	120
Figure 14:	Myelinated and Demyelinated Axons	131
Figure 15:	Food for Brain Health	173
Figure 16:	The Stages of Sleep	204
Figure 17:	Circadian Rhythm	205
Figure 18:	SMART Targets	222
Figure 19:	Factors Contributing to Resilience	232

Preface

I have wrestled with this book for the best part of a decade. Its origins go back to when I was doing research for my doctorate in University College Cork, Ireland. At the time, my interest was in exploring how adults learn and especially how they experience returning to formal education. To this end, my thesis explored the philosophy of andragogy (how adults learn), which gradually led me to the world of neuroscience. Recent years have seen great interest in the link between neuroscience and education and the more I discovered about the brain the more I was surprised at not realising the obvious: that learning is a brain function, and trying to understand it is futile unless we first understand the brain. I became fascinated by the new and innovative research, with its insights into how the brain worked and how people learn.

Later when I was approaching formal retirement from the University I was asked by the Department of Adult and Continuing Education to know if I would be interested in delivering a short course on a topic of my choice. The purpose of the short courses in UCC was to engage with the wider community offering courses dealing with a wide variety of new ideas and concept. These courses also allowed university staff to bring their most recent research into the public domain. Coincidently at this time there was great impetus generated into brain research by new developments in technology. Indeed, the American Congress declared the 1990s The Decade of the Brain.

As a result of this request, I proposed to run a six-week course on brain health. This has been an incredible journey for me because it has been running twice each year for the past decade. During and after the course, many adult learners have expressed their absolute astonishment with the information I was presenting about the brain, and its significance to life, education and relationships. The constant refrain was twofold: firstly, why is this information not more readily available to teachers, students and young people, and, secondly, why don't you write a book about it!

Around this time, I also worked with Dr Aimie Brennan on a number of research projects. Aimie is a champion of teacher education, and a firm believer in the privilege and power that education can bestow. During our work together, we had many conversations about the links between neuroscience and education, and the importance of evidence-based teaching and learning. This later led to the idea of both of us combining our interests and skills to explore how neuroscience and education could learn from each other.

Together, we have spent the last four years delving into this new world of educational neuroscience, which we found exciting and one which we wanted to share with the wider educational community. We hope that this book will inspire other younger scholars to take up the challenge of continuing what we have started, by engaging in further collaboration between neuroscientists and educators so that eventually both will learn and prosper from the blending of science and practice.

Acknowledgements

We are grateful to the many people whose insights have contributed to the researching and writing of this book. In particular, we would like to thank all of our students and graduates who continually provide valuable feedback and ask important questions, all of which inspired the research for the book.

To the pioneering scholars who have helped to establish and build educational neuroscience as a new field of enquiry through their openness to transdisciplinarity, particularly Professor Ian Robertson, Dr Harris Eyre, Professor Usha Goswami, Dr William H. Kitchen whose voices are included as vignettes throughout the book. The views and opinions expressed are, of course, those of the authors, alone.

Many thanks to our partners for their patience, perseverance and unwavering support, to our families and friends, particularly Anthony Hartnett for all the conversations, arguments, insightful corrections and incredible support throughout this journey. A final word of thanks to all the teachers who educate with passion, patience and the pursuit of knowledge.

Introduction

> 'The greatest voyage of self-discovery is not to seek new landscapes but to find new eyes'.
>
> (Marcel Proust 1855: 261)

So, why the title, *Brain Health for Learning: what neuroscience brings to Education*. Well, at one level, it tries to capture a rather simple idea, that in neuroscience we learn how the brain and neural system function. In education, we encounter the art and mystery of learning. If the brain is about thinking, and education is about learning, then surely both have something powerful to say about how teaching and learning happen. The emerging field of educational neuroscience attempts to incorporate insights from neuroscience into education and it is an exciting new area, now that we know much more about how the brain works and learns. Equally, we get to clarify how learning is related to brain structures and its growth and enhancement.

In recent years, brain scientists have drawn our attention to some remarkable discoveries on how the brain has the capacity to change. Indeed, studies using magnetic resonance imaging (MRI) brain scanning technology have shown that new neural cells and neural pathways are generated throughout one's life. This ability of the brain to reorganise itself by forming new neural connections is known as neuroplasticity. What we now realise is that the way the neurons are wired together can be dramatically changed because of experience, learning and behaviour. Everything related to neuroplasticity is revolutionary, as it has opened new doors to thinking with new eyes about health, education and lifestyle. We each have much more control over our own brain than perhaps we realise. In this context, one of the most profound insights of contemporary neuroscience is that the brain has enormous capacity for self-healing and regeneration.

Simultaneously, educational neuroscience has impacted on educational practice by generating applied interdisciplinary research that provides a new way of considering learning. Although there is still a great deal of mystery surrounding the brain, we now have sufficient information from neuroscience, psychology and education that we can harness this information to both understand and optimise our brain health and performance. The quality of education has been positively changed by the rapid development of science. The fact that we now know that the brain is malleable, and that new neural pathways can develop throughout the lifecycle, has opened the door to new kinds of thinking and performance.

Studies of development psychology, cognitive psychology, pedagogical science and neuroscience have all converged on a new understanding of the workings of the brain. Recent insights from these different perspectives highlight that learning changes the physical structure of the brain, organising and reorganising so that, different parts of the brain may be ready to learn at different stages of development (Zull 2002). During its development, the wiring of the brain is created through the formation of synapses, which are the junctions between the neurons through which information passes. Understanding how the brain works has led educational neuroscience to some interesting revelations in understanding learning in a deeper sense, with the subsequent debate placing student learning at the very centre of the educational enterprise.

Modern neuroscience has built a sophisticated picture of how our brain works. This knowledge offers a new paradigm and a dynamic developmental model for brain growth and change throughout the lifespan. One example of this is the development of educational neuroscience as a new perspective within the curriculum in some third-level courses, ranging from childcare, primary, secondary, adult continuing education and gerontology. Indeed, educational neuroscience, not only puts a spotlight on what is going on inside our brains, it also shows us how to improve it (Watagodakumbura 2017). In fact, better understanding the capacity of the brain and the mental processes involved in learning can help us to harness our own power and support us in our individual learning journey.

The role of the brain in learning is not a new topic and, understandably, there is a lot of enthusiasm for the science of teaching, informed by developments in educational neuroscience. Much of the brain research that has occurred over the last twenty years has focused on brain dysfunctions and findings have not always been viewed as highly applicable to typical learners (Schunk 2012). In Ireland, in March 2021, there was a national campaign called *Love your Brain* which aimed to promote greater understanding of the brain and brain conditions. In July 2021, clinicians from all over Ireland came together to support *World Brain Day*.

The move towards understanding the learning process calls for a paradigm shift at the philosophical and policy level. For some policy makers, our learning and education should take us on a path to higher levels of human development and self-actualisation. When we progress in human development through lifelong learning, we enhance our consciousness, metacognition (thinking about thinking) and wisdom. In effect, we become conscious of our mental and physical well-being and pursue a healthy lifestyle by engaging in more reflective or metacognitive approaches to learning.

An Association of American Colleges report in 1985 recommended that the central theme of any curriculum should be to teach students 'how to learn' (Wirth and Perkins 2013: 5). For all the reasons given above, and for many others, the focus of education is shifting from 'teaching' to 'learning to learn'. As John Naisbitt writes, 'In a world that is constantly changing, there is no one subject or set of subjects that will serve you for the foreseeable future, let alone for the rest of your life. The most important skill to acquire now is learning how to learn' (1988: 5). Furthermore, we now see a greater awareness of the need for respect for people different from ourselves as well as recognising people's ability to change and the ongoing desire for empathy, resilience and disposition for lifelong learning.

Lifelong learning consistently contributes to our mental well-being in a variety of fashions. Lifelong learning does not merely mean to be enrolled in an educational institution to engage in learning. It is also about how we use our brains by engaging in mental activities that help us grow our integrated neural networks of knowledge. What is important here is that neuroscience reminds us that, by paying attention and being reflective of our experiences, we learn more deliberately and in a more focused way. This is especially relevant to the volume of information and disinformation we encounter daily.

What Is This Book About?

This book is an original contribution to the field of educational neuroscience. We argue that all adults should pay more attention and become more reflective of our experiences so that we can learn more deliberately and in a focused way. This book addresses the functioning of the brain, the architecture of the brain, neuroplasticity, learning, and how to develop and maintain a healthy brain throughout life. Learning to learn is about individuals understanding how they learn and thus developing their capacity to learn.

There have been calls for new kinds of learning from many different parts of society, with the educational system transformed by the many challenges presented by the COVID-19 pandemic from 2019 onward, as well as by social media and Artificial Intelligence challenges. Educators from across all sectors are much more aware of the need for enhanced critical competencies over and above specific academic domains, including personal responsibility, the ability to act in a principled, ethical fashion, skills in oral and written communication, interpersonal and problem solving abilities.

We argue that by better understanding the way in which the brain operates, we can each develop a nuanced understanding of learning processes in both how to prepare our brains for deep learning and how to employ strategies that use brain science to optimise learning. In this book, we draw on research by neuroscientists, educationalists and psychologists in order to unpack key learning theories such as spaced learning, emotional connection, transfer learning, metacognition, scaffolding and interleaved learning, all of which can help explain how learning occurs from a neuroscience perspective.

Who Is This Book For?

While there are a multitude of texts aimed at those in the medical profession, and various resources online for those interested in brain health,

Introduction 5

this current book provides an accessible introduction to the connection between the brain and learning for undergraduate students, teachers and lifelong learners. In particular, we have written this book for learners who are transitioning or returning to education, those training to work with children and young people, and those interested in understanding more about their own learning. This book may also be useful for those planning to work in formal and informal education, for workers in community services dealing with the elderly and for those looking to make small lifestyle changes that will have long-term benefits for their own brain health and learning.

We are familiar with the idea that certain lifestyles are unhealthy for us. We know we should not smoke or abuse alcohol; we are aware that sitting at a desk for 8 hours per day is not a great idea; and that we should avoid eating lots of sugar. In short, we need our brain and our brain needs us. Because this book deals with issues of health promotion and social engagement, it will also be of interest to anyone who is interested in learning about and supporting brain health in themselves and others. We have developed a helpful acronym LIFESTYLE to highlight how everyday choices can enhance brain health and learning. We hope that this book will be a useful teaching resource for any programmes that deal with the transition to higher education, for teaching and learning in the twenty-first century and general teaching and learning skills for undergraduate students.

Features of the Book

The following features are included throughout the book to connect research, policy and practice in the field of educational neuroscience:

Expert Insights

Short 'expert insights' are included throughout the book to provide fresh perspectives on contemporary neuroscience research and its implications for brain health and learning. We have included the voices of experts with alternative perspectives on neuroscience, educational neuroscience, brain health and philosophy, from national and international contexts. In recorded interviews, we asked experts for their views on the key learning from recent breakthroughs in neuroscience research and why they think educational neuroscience should be more accessible. Expert insights include direct quotes and current publications from the following researchers:

- Professor Ian Robertson, Professor of Psychology at Trinity College Dublin, who was the founding Director of Trinity College Institute of Neuroscience
- Dr Harris Eyre, fellow in brain health at the Center for Health and Biosciences and Co-Founder of the PRODEO Institute, and advisor to the Organisation for Economic Co-operation and Development (OECD).
- Professor Usha Goswami, Director for the Centre for Neuroscience in Education and Professor of Cognitive Developmental Neuroscience and a Fellow of St John's College, University of Cambridge.
- Dr William H. Kitchen, lecturer at Stranmillis University College, Belfast and author of *Philosophical Reflection on Neuroscience and Education* (2017).
- Dr Yvonne Nolan, Vice Dean of Graduate Studies, College of Medicine and Health, UCC and colleagues, Dr Eithne Hunt and Dr Samantha Dockray, who developed 'Brainpower', focusing on brain development for staff across further and higher education.

Case Studies of Educational Neuroscience

Dispersed throughout the chapters in this current book, case studies are examples of on-going research that indicate the diversity and range of research topics being investigated. The purpose of the case studies is to provide insight into innovative and contemporary research in the field of educational neuroscience, and to provide additional resources for readers who wish to explore specific aspects of theory and practice.

Introduction

Overview of the Book

Section 1

The first section of the book provides a foundation to the biology of the brain and the historical development of modern neuroscience as a discipline. The section provides the reader with a context for understanding the functions and architecture of the brain and how it contributes to learning. It also introduces the reader to the terminology and science behind how we learn, particularly the role of neurons, neuroplasticity and habits.

Section 2

The second section provides an overview of the development of educational neuroscience and its role as a discipline in helping us understand the link between learning and brain health. This section provides the reader with an understanding of the neural mechanisms and how they are shaped by the learning process.

Section 3

The third section addresses the 'Lifestyle' model. Here, we address key concepts from neuroscience research to help learners explore how everyday choices can enhance brain health for learning. This model provides a framework which utilises research from neuroscience across a range of human activities. The model is an acronym for the following:

- L – Make learning novel and complex
- I – Interpersonal relationship – stay connected
- F – Food and our second brain
- E – Exercise and neurogenesis
- S – Sleep and consolidation of learning
- T – Targets for dopamine

- Y – Yearning for meaning
- L – Liquids the elixir of life
- E – Empathy and mirror neurons.

This framework provides the reader with knowledge and techniques (tools) that support them to make the connection between neuroscience and learning.

References

Lucas, B. and Greany, T. (2010). Learning to Learn: Setting the Agenda for Schools in the 21st Century. In Amalathas, E. *Learning to Learn in Further Education: A literature review of effective practice in England and abroad.* London: Centre for Better Teaching (CfBT Educational Trust). Network Education Press.

Naisbitt, J. and Aburdene, P. (1985). *Re-inventing the Corporation.* New York: Little Brown & Co.

Naisbitt, J. and Aburdene, P. (1990). *Megatrends 2000.* New York: William Morrow.

Proust, M. (1855). *Remembrance of Times Past*, Vol. 3.

Schunk, Dale H. (2016). *Learning Theories: An Educational Perspective* (6th edn). Boston: Pearson Press.

Watagodakumbura, C. (2017). *Programming the Brain: Educational Neuroscience Perspective.* SC, Charleston, South Carolina: Create Space Independent Publishing.

Wirth, K. and Perkins, D. (2013). *Learning to Learn.* Macalester Course Materials. Montana State University. Available from: <https://www.montana.edu/rmaher/barrier_courses/Learning%20to%20Learn%20Wirth.pdf>

Zull, James E. (2011). *The Art of Changing the Brain: Enriching the Practice of Teaching by Exploring the Biology of Learning.* Sterling, Virginia: Stylus Publishing.

SECTION I

CHAPTER 1

Meet Your Brain

'We never stop investigating. We are never satisfied that we know enough to get by. Every question we answer leads on to another question. This has become the greatest survival trick of our species'.

(Morris 1967: 107)

Horns blare! A child, who has not seen a car speed past, leaps back to the safety of the footpath. A waiter is balancing a tray piled with dishes with one hand. It wobbles precariously and looks like the whole stack will come down, but he swiftly, instinctively saves it. A novice tennis player follows a clear step-by-step guide given by a coach in how to serve properly, thus learning the importance of a routine. A graphic designer wrestles for days with a problem that seems to defy solution. And then ... then, the answer comes to her out of the blue while thinking about something entirely different.

The above are just some examples of the routine functioning and capability of the human brain. Its presence in the body makes each of us constantly responsive to the environment through the senses, an extraordinary super-powered processor capable of boundless and interconnected thoughts. Our brain is the basis for who we are, our intellect and our personality. Without it, we would be deprived of such amenities of life as the enjoyment of food, or music, or the colour of a painting, or the pressure of a friendly handshake. We depend on our brain when we interact with other people, from subtle emotional responses to language and to reading cultural customs. From the moment we are born to the moment we die, this communication network controls our every thought, step, emotion and impression. We depend on our brain for our capacity to learn, recall, connect, remember and grow.

And yet ... it is still possible to go through life oblivious of the capabilities and potential of the brain.

In this chapter we will examine how the secrets of the human brain have been gradually revealed. Building on the early work of anatomists and recent research in the neurosciences, we will address fundamental questions about how the brain works. By describing the underlying structures and mechanisms of the brain, we highlight the way in which the brain governs aspects of how we learn.

The notion that we only use 10 per cent of our brains is a myth. According to Alistair Smith in his book *The Brain's Behind It*, this myth 'probably survives within the self-improvement industry by misinterpreting early researchers who said they only know at most 10 per cent of how the brain functions' (2002: 22). We use all of our brain, just not at the same time. Considering its importance, the brain is surprisingly small, and, in many cases, it is just over six inches from front to back, a simple fact that makes it hard to imagine how such a small mass of tissue can be the source of all aspects of our humanity. In this context scientists are still mystified by consciousness and the whole world of the subconscious mind.

For all its formidable role, the brain is not visually attractive. To hold it in the palm of your hand, it is a wet, gooey, fatty piece of rather crinkled and whitish flesh, having the slightly rubbery feel of a large mushroom. It does not pump like the heart, expand or shrink like the lungs, or secrete visible material like the bladder. If we sliced off the top of a skull and peered in, we would not see much happening at all. It is an incredibly fragile and sensitive organ. In fact, if the brain does not receive blood for 10 seconds, we would become unconscious.

The average weight of an adult brain is 1,200–1,400 g (about 3lbs), and according to the Indian neurologist, Dr Sudhir Shah, has a capacity for generating 30,000 thoughts a day (Shah and Shah 2017: 5). Interestingly, it uses around 20 per cent of our body's energy and 50 per cent of our daily oxygen intake. Approximately 73 per cent of our brain is water. In addition, our brain is always active, never shutting down, even while we sleep (Costandi 2016). In fact, during sleep, the brain uses its own housekeeping processes to clean out the toxins accumulated during the day.

Sitting inside the skull, the brain essentially floats, suspended by its own special fluid called cerebrospinal fluid (CSF). Cerebrospinal fluid both surrounds and cushions the brain and spinal cord. Its major function is to act as a shock-absorbing fluid to prevent the brain from bumping against the interior of the hard skull when the head is subjected to sudden, jarring movements. Additionally, the CSF plays a pivotal role in 'washing' the brain, particularly during sleep, by removing waste (Huffington 2016). The glymphatic system is a recently discovered waste clearance system which flushes toxins out of the brain using CSF (Jessen et al. 2015). This process of waste elimination will be discussed in more detail in Chapter 12.

Learning from Ancient Civilisations

The human brain began to evolve over 500 million years ago, but it is only in the last 500 years that scientists began to discover its secrets. Even more amazing is the fact that 95 per cent of what we know about the brain and how it works was discovered within the last sixty years. A Congressional Resolution signed by George Bush designated the 1990s as the Decade of the Brain. In a Presidential proclamation, he wished to 'to enhance public awareness of the benefits to be derived from brain research' (1990). A decade later, neuroscientist, Antonio Damasio, wrote 'more may have been learned about the brain and the mind in the 1990s – the so-called Decade of the Brain – than during the entire previous history of psychology and neuroscience' (1999: 4).

To the naked eye, there is very little to see regarding what is really going on in the human brain. Perhaps it is not so surprising then that it has taken humans such a long time to understand its complex nature. Most old civilisations regarded the heart, and not the brain, as the centre of the soul. Ancient Egyptians, when embalming human bodies, religiously preserved the heart, but destroyed the brain. But it was an Egyptian surgeon who left the first written descriptions that gave evidence of some basic insight into understanding the brain.

The Edwin Smith Papyrus, named after the nineteenth century American discoverer, and dated to 1700 BCE, reveals that the Egyptians too recognised a relationship between the heart and blood vessels some 3,300 years before William Harvey issued his noteworthy verdict on the circulatory system. The Egyptians' main source of anatomical education was the embalming process. In preparing their dead to be mummified, they had to remove the more perishable parts of the body, such as the brain, lungs and intestines, to preserve the rest. In recent years, X-rays of mummies have confirmed the deftness of these manipulations.

However, embalming was a process of preservation, not a discipline for studying the human body. Archaeologists have turned up clay tablets with hieroglyphs depicting the stomach, liver, womb, as well as symbols representing the brain. The hieroglyphics go on to describe how a person may not be able to move one limb after severe head injuries on one side of the body or the loss of speech resulting from injuries to the temple, several thousand years before Dr Paul Pierre Broca described the speech centre in the 1860s.

The Chinese prescribed all manner of rules for healthful living, such as regular exercise, avoidance of improperly cooked foods, and the use of mouthwash; no less a sage than Confucius cautioned, 'Diseases enter by the mouth'. He also intuitively knew that the brain was the seat of learning. To this day, his emphasis on the importance of 'reflection' in teaching and learning is captured in his advice, 'To learn without thinking is blindness, to think without learning is idleness' (Cements 2007: 44). Thousands of years elapsed before the brain, rather than the heart, was universally recognised as the most important organ in the body.

Early Personalities

During the golden age of Greece in the fifth century BCE, individual personalities began to leave a deeper imprint on health matters. The first was Hippocrates (c. 460–375 BCE) whose writings became akin to a bible on medical practice in the early western world. He recognised the unique nature of the brain: from the brain only arise our pleasures, joys, laughter, as well as our sorrows, pains, grief and tears. His extensive collections of writing, as well as advice to physicians, has been epitomised in the Hippocratic Oath, still taken by students attending medical school today. While his contemporaries, including Aristotle, wrongly believed that the brain was to be found in the heart, Hippocrates correctly argued that the brain is the seat of thought, sensation, emotion and cognition.

The influence of the next towering figure in the history of medicine was the 'gospel' according to Galen, a second century BCE Roman physician whose theories animated medicine for 1,400 years until the late Middle-Ages and beyond. Indeed, through his medical writings he gave far greater precision to existing knowledge of anatomy and physiology. Many of his descriptions on the movement of muscles and the intricate workings of nerves appear in medical textbooks to this day. Galen observed the effects of brain injury in Roman gladiators and came to the correct conclusion that the brain controls our movements.

However, when it came to understanding brain science, Galen ignored the solid brain tissue in favour of the fluid-filled cavities, or ventricles. Galen believed the human brain to have three ventricles, with each one responsible for a different mental faculty: imagination, reason and memory. According to his theory, the brain controlled our body's activities by pumping fluid from the ventricles through nerves to other organs. Yet, the idea had a major flaw: a fluid could not move quickly enough to explain the speed of our reactions. However, such was Galen's reputation and authority that this idea cast a long shadow over our understanding of the brain, and subsequently fluid theory of the brain dominated medical practice for hundreds of years.

With the restless new spirit of inquiry that began to pervade the western world as the Renaissance approached in the fifteenth century, Galen's extended reign came to its slow end. His dethronement was bitterly fought, as his teaching had both Church and academic backing (Singer 1962).

In the seventeenth century, English doctor Thomas Willis presumed to challenge Galen's supremacy, and prevailed. Willis argued that the secret to the functioning of the brain lay in the solid cerebral tissues, not the ventricles. However, it was not until the nineteenth century that German physiologist, Emil du Bois-Reymond, confirmed that nerves and muscles themselves generate electrical impulses. This work paved the way for the modern era of neuroscience but before this could happen brain scientists had to wait a few hundred years for a certain type of technology to develop to assist them view the brain in new ways.

Spanish anatomist Ramon y Cajal (1852–1934), identified neurons as the building blocks of the brain. He found them to have a diversity of form that is not found in the cells of other organs. Most surprisingly, he noted that insect neurons matched and sometimes exceeded the complexity of human brain cells. His research suggested that our abilities depend on the way our neurons are connected, not on any special feature of the cells themselves. Cajal's 'connectionist' view opened the door to a new way of thinking about learning and information processing in the brain which still dominates today (see Chapter 3 on neurons).

For thousands of years, physicians relied on studying the human body, armed only with the naked eye, often relying on intuition to treat disease and illness. However, in sixteenth century Europe, new developments taking place outside of medicine would change our understanding of the human body forever. Technology was radically changing all aspects of life, society and medicine. Simple magnifying lenses had been known as far back as the first century, with the Romans using them for reading and starting fires. But in 1590, Hans and Zacharias Janssen, a father and son team of Dutch spectacle makers, devised a microscope. As the first crude instruments gave way to finer models, they revealed an assortment of incredible wonders, most notably to the inquisitive eyes of Anton van Leeuwenhoek, a clothes merchant from Delft. Fascinated by the potential of the telescope, Leeuwenhoek spent his free time scrutinising any substance that came to hand – pond water, milk, blood, and meat – and discovered they were teeming with bacteria.

Leeuwenhoek endeared himself to later researchers by documenting everything he observed. Over the course of fifty years (1673–1723), he bombarded the Royal Society in London with more than 200 letters outlining his scientific observations (Hoole 1800). Soon his fame spread, and his study of cells opened the door to a completely new conceptualisation of the architecture of all living organisms, including the human body. It did not dawn on him, however, that the body might be made up *entirely* of cells, a fact that remained undiscovered until a century later when Theodor Schwann (1839) confirmed that all living matter is composed of cells similar in structure and function. Now, scientists could get to the root of the matter: the cellular nature of all living things.

The Birth of Neuroscience

Medical science during the latter half of the twentieth century split into the study of physical anatomy which examined what could be seen of the body with the unaided eye, and microscopic anatomy which explored that which had up to then been invisible, and thus only speculative. As we will see, this had a profound impact on brain research.

Before we were able to look inside the brain, phrenologists tried to understand behaviour and claimed to reveal our personality traits by measuring the bumps on the skull (Greenfield 1997: 11). German doctor, Franz Joseph Gall, announced in the 1790s that he could tell a person's character by examining their skull for lumps. The apparatus that Gall used to make his analysis was a kind of hat. When placed on the skull, movable pins were displaced by the bumps on the surface of the skull so that they were pushed upwards to pierce through paper. The particular pattern of perforations in the paper thus gave a somewhat unique readout of an individual's character.

Gall's theory was that each of the bumps on the skull represented a certain character trait (see Figure 1). He divided the brain into thirty-two different personality traits, ranging from cleverness, vanity, cruelty, firmness, cautiousness, vitality, friendliness, inquisitiveness, agreeableness and so on. He argued that he could identify whether a person was a loving parent, a devout Christian, or a calculating murderer by reading their skull. Indeed, phrenology was responsible for some thoroughly laughable claims, such as the idea that Eastern peoples were less warlike because they had smaller heads. The iconic poster depicting the regions of the brain like a coloured map is frequently found on modern day post cards or posters.

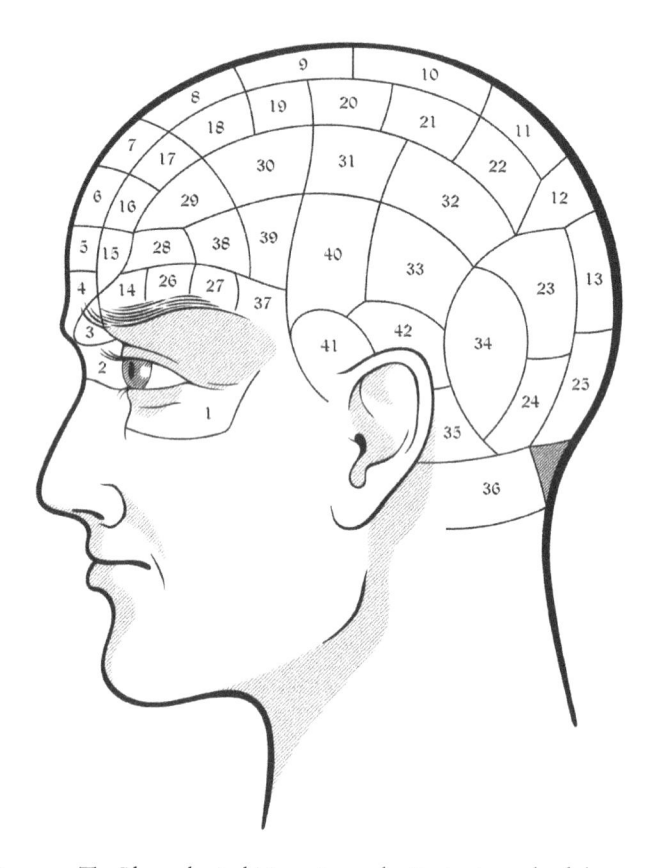

Figure 1: The Phrenological Map – Image by Croisy/<stock.adobe.com>

For a time, phrenology grew in popularity among the emerging middle classes as a form of entertainment in the newly formed coffee parlours throughout European capitals. The Scottish scientist, John Gordon (1815), dismissed it as 'a piece of thorough quackery from beginning to end' (cited in Gieryn 1999). However, while the method used by phrenologists may be similar to people today reading their star signs to guide their behaviour, the idea that different parts of the brain controlled or managed different functions in the brain does have some validity. As we will see, phrenology perhaps opened the door to a more accurate architecture of the human brain.

Modern neuroscience can be traced back to the mid-1850s, when researchers first determined that the way to investigate how the brain works was to examine the behaviour of individuals experiencing brain damage. One of the earliest, and best-known case studies in neuroscience is that of a railway worker named Phineas Gage. When he went to work on the morning of 13 September 1848, the 25-year-old had no idea that he was going to be immortalised in medical and psychology history for years to come. He was a railway worker living in Vermont in the US. One of his jobs was to set off explosive charges in large rocks to break them into smaller pieces. To do this, he used a tamping iron rod to pad in the explosives. On this day, Gage inadvertently caused an explosion that propelled a metre-long iron rod through his skull at great speed. The rod entered his skull below the left eye, continuing through the front left region of his cortex, and soaring out the top of his head (Fleischmann 1948). The rod was later recovered about 30 feet from the scene of the accident, 'smeared with blood and brain'. Remarkably, Gage survived!

The iron rod had destroyed much of Gage's left-frontal cortex as it passed through his head. This affected his decision-making abilities, and made him lose many of his social inhibitions. According to popular accounts, the result was a profound change in his personality. Those who knew him reported that he had changed from a mild-mannered young man to a hot-tempered, impulsive individual, who seemed like a different person, and that he was 'no longer Gage'. Within a few months, he was let go from his job in the rail company and went on to hold down several menial jobs. Despite living for eleven more years, he was never again to become his former self. Towards the end of his life in 1860, he toured with a circus, in which he was exhibited, together with the iron rod that impaled his head.

The case of Gage has become something of a legend as physicians later tried to reconstruct his skull in order to determine the extent of the damage, with psychologists speculating on what may have brought about his personality changes. For instance, the change in his personality was attributed to the damage to his frontal lobes, that area of the brain just above the eyes which we now know is responsible for making decisions, problem solving, and emotional expression. In this sense, we can say that the Gage incident sparked the beginning of modern scientific neuroscience because his injuries stimulated future scientists to learn more about the human brain when things go wrong.

Two ideas informed by Gage's accident are still important today. Firstly, it was discovered that the brain has separate parts, each specialised to do specific things. Abilities such as movement, language, decision-making and social cognition came to be associated with the frontal lobes, whereas memory, perception, and intelligence are spread throughout many different parts, thus making them more challenging to know how they work. This idea, through the dedication and innovation of many creative people over many years, eventually led to mapping the architecture of the human brain (Al-Chalabi et al. 2018).

Secondly, Gage's situation led scientists to question their assumption that the brain was 'hardwired', that people born with mental limitations or learning disorders were destined to remain so. Scientists eventually discovered the existence of neuron brain cells which can 'switch' on genes that change mental structure as learning occurs. Gage's accident thus helped scientists to investigate where different functions may reside, as well as the possibility that the brain may 'change' itself over time after an accident.

Nowadays, however there are many advanced brain imaging technologies revealing how the brain works, the role of its constituent parts, and what happens when these parts malfunction. In recent years, we have clearly arrived at a point where we understand something vital about humanity: that the brain is the key organ responsible for who we are. But, as we have illustrated, this was by no means something we had always understood. Indeed, the key process, which developed with improved scientific measurement from the middle of the nineteenth century onwards, was brain mapping.

The process of mapping the architecture of the brain has provided brain scientists with some of its most crucial information. Thanks to pioneers in this field, we now know that there are regions of the brain responsible, at least in part, for movement, sensation and vision, and others for such things as language, speech, processing information, learning and intuition. This complexity came with a cost however. Not only do all these systems have to be developed and interconnected, but they also have to stay balanced and properly integrated for optimal performance. It is in this context that we have now to consider, in the next chapter, why the architecture of brain health and learning are important.

Conclusion

In this chapter, we have described the powerful function of the human brain. We examined the historical development of brain science, described the major paradigm shifts and conceptual revolutions that have taken place in science while tracing the contributions of key researchers and important discoveries that led to the birth of neuroscience as a distinct field. By better understanding the historical context, and our current place within it, we can appreciate how neuroscience may progress as knowledge is accumulated and technology advances into the future.

References

Al-Chalabi, A., Turner, M. and Delamont, R. (2018). *The Brain: Beginners Guide*. London: One world Publishing.
Bush, G. (17 July 1990). *Presidential Proclamation 6158*. Filed with the Office of the Federal Register, 12:11 p.m., 18 July 1990. Available from: <https://www.loc.gov/loc/brain/proclaim.html>
Calvin, W. (1997). *How Brains Think: Evolving Intelligence, Then and Now*. London: Weidenfeld and Nicolson.
Confucious. (c.551 BC). *The Analects-Confucius (551–479 BCE)*. Translated by J. Legge. China.
Costandi, M. (2013). *The Human Brain: 50 Ideas You Really Need to Know*. London: Quercus Books.
Damasio, A. (1999). How the Brain Creates the Mind. *Scientific American Magazine* (Special Issue), 281(6), 112.
Damasio, A. (2002). How the Brain Creates the Mind. *Science America (Special Issue)*, 12(1), 4.
Eagleman, D. (2015). *The Brain: The Story of You*. Edinburgh: Canongate Books.

Fleischmann, J. (1948). *Phineas Gage: A Gruesome But True Story about Brain Science*. Boston: Houghton Mifflin.
Foder, A. (1983). *Modularity of Mind: An Essay on Faculty Psychology* (p. 2). Cambridge, MA. MIT Press.
Gieryn, T. (1999). *Cultural Boundaries of Science: Credibility on the Line*. Chicago: University of Chicago Press.
Greenfield, S. (1997). *The Human Brain: Guided Tour*. London: Orion Books.
Greenfield, S. (1998). *The Human Brain: A Guided Tour*. London: Orion Books.
Hajar, R. (2012). The Air of History: Early Medicine of Galen (Part 1). *Heart Views*, 13(3), 120-128.
Harvey, W. (1889). *On the Motion of the Heart and Blood in Animals*. London: George Bell and Sons.
Hoole, S. (1800). *The Select Works of Anthony van Leeuwenhoek*. London: Sidney.
Huffington, A. (2016). *The Sleep Revolution: Transforming Your Life, One Night at a Time*. London: Ebury Publishing.
Jessen, N. A., Munk, A. S., Lundgaard, I. and Nedergaard, M. (2015). The Glymphatic System: A Beginner's Guide. *Neurochemistry Research*, 40(12), 2583–2599.
Koch, C. and Marcus, G. (2016). 'Neuroscience in 2064: A Look at the Last Century'. In G. Marcus and J. Freeman (Eds), *The Future of the Brain: Essays by the World's Leading Neuroscientists*. USA: Princeton University Press.
Loveday, C. (2016). *The Secret World of the Brain: What It Does. How It Works and How It Affects Behaviour*. London: Sevenoaks.
Morris, D. (1967). *The Naked Ape: A Zoologist's Study of the Human Animal*. UK: Jonathan Cape Publishing.
Mukherjee, S. (2022). *The Song of the Cell: An Exploration of Medicine and the New Human*. London: Penguin Random House.
Shah, S. V. and Shah, H. S. (2017). *BRAIN and Neurological Disorders: A Simplified Health Education Guide*. New Delhi: Jaypee-The Health Sciences Publisher.
Singer, C. and Underwood, E. (1962). *A Short History of Medicine*. UK: Oxford University Press.
Smith, A. (2002). *The Brain's Behind It: New Knowledge about the Brain and Learning*. Stanford: Bloombury Publishing.
Smith, E. (n.d.). Papyrus: Egyptian Medical Textbook of Surgery and Anatomical Observations. *Encyclopaedia Britannica* (online edition). Retrieved 20 October 2021.

CHAPTER 2

The Architecture of the Brain

> 'The most staggering thing about our brain is not what it is but what it can become'.
>
> (Gilbert 2008: viii)

A necessary step in learning to live in an unfamiliar city is learning the names and location of landmarks, major neighbourhoods and districts. Those who possess this information can easily communicate the exact location of any destination in the city. This chapter introduces the various neighbourhoods and districts of the brain for much the same reasons. We will describe the basic functioning of the brain and its architecture. Although such an introduction is inevitably somewhat technical, it is important to have an overall idea of the brain's complexity as well as its anatomical regions and distinctive features to appreciate how it facilitates human learning.

The human brain was shaped over millions of years by sequential adaptation in response to ever changing environmental demands. Over time, brains grew in size and complexity, whereby old structures were conserved and new structures emerged. As we evolved into social beings, our brains became incredibly sensitive to our social world. This mixture of adaptation and innovation resulted in an amazingly complex brain capable of everything from monitoring our breathing to delving into the very origins of the cosmos. However, as noted in the last chapter, the complexity came with a cost. Not only do all these parts have to be developed and interconnected, but they also have to stay balanced and properly integrated for optimal performance.

A challenge faced by early brain scientists was that there are no obvious moving parts within the brain, as it does not operate mechanically as our hearts and lungs do. Early anatomists sought to explain what they observed by dividing the brain parts by location and identifying three main structures: the forebrain, midbrain and hindbrain. A later perspective, proposed by Paul MacLean in the 1960s, described the triune brain according to stages of evolution: reptilian, paleo-mammalian and mammalian. According to this evolutionary story, the human brain ended up with three layers: one for surviving, one for feeling and one for thinking (Maclean 1990).

To form a basic foundation in knowing how the brain works, we will examine the major outer parts of the brain, including the two hemispheres, the four lobes and the cerebellum. We will then look at the interior of the brain and divide it into parts on the basis of their general functions: the brain stem and the limbic system, including the basal ganglia, and the cerebrum which sits on the surface of the brain. The brain's physical structure broadly reflects its mental organisation. In general, higher mental processes occur in the upper regions, while its lower regions take care of basic life support.

Imagine you are looking at a brain in your hand. The first thing you notice is that it resembles a head of cauliflower. From the outside, its most distinguishing features are its folds, which allow the covering to maximise its surface area. If the folds were laid out flat, it would be the size of a daily newspaper. The folds (or wrinkles) are part of the cerebral cortex (Latin for bark), the brain's outer covering. In size, the brain looks like a large grapefruit. You would also observe that the brain is made up of distinct regions that fold and interlock around each other. The brain has a pink-grey hue from the many tiny blood vessels that provide it with oxygen and nutrients (Greenfield 1998). If you look down from the top, you would see that the brain has two clear halves, called hemispheres, similar to a giant wall-nut. The hemispheres are one of the most striking structures of the brain, sitting on a central core or base.

Neuroscientists have divided the brain into numerous subdivisions in order to make locating parts easier (Clancy 2018). The first subdivision is the left and right hemispheres, which anatomically look almost identical. The way we name the different areas is the same for both sides of the brain. The cerebrum is divided down the middle into a right hemisphere and a left hemisphere. The left and right side of the cortex are broadly similar in

The Architecture of the Brain

shape, and most cortical areas are also replicated on both sides. The two sides of the hemisphere halves make up the largest portion of the cerebrum and are clearly distinguished by a longitudinal fissure, which is the deep groove in the centre of the brain. For reasons that are still unclear, the connections are crossed, the right half of the brain controls the sensory and movement functions of the left half of the body and vice versa.

The scrunched-up outer layer of both hemispheres is one of its most recognisable features.

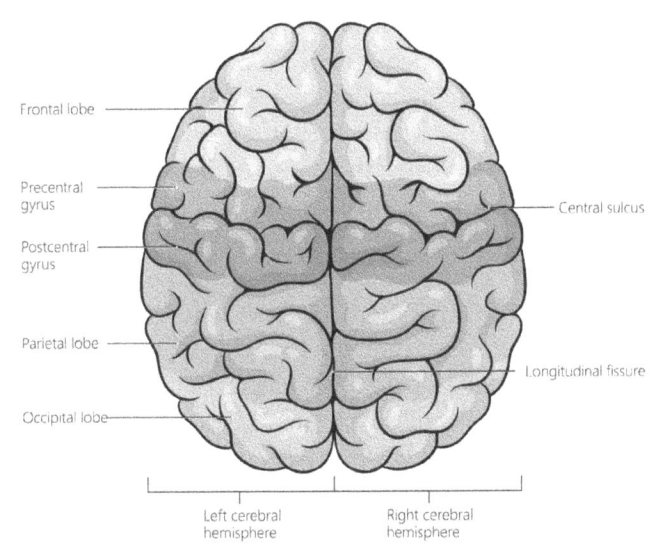

Figure 2: Human Brain Top View – Image by *Oleksandr Pokusai*/<stock.adobe.com>

The convolutions have the effect of increasing the capacity for neural connections in the cerebral cortex without increasing the overall volume of the brain. Each hemisphere contains a significant part of the cerebral cortex, which is generally considered to be the seat of higher-level cognitive functions, as well as sensory and motor control. Most of the cerebral cortex is called the neocortex (the 'new' cortex) because it is the most recently developed part from an evolutionary perspective. It is this part that is believed to separate human beings from other species, since it is the seat of our two most distinguishing characteristics: our capacity to learn and the ability to imagine the minds of others (Alderseya Williams

2014). We will return to the combination of these unique characteristics in later chapters.

The two hemispheres have given rise, in popular literature, to the notion of the 'male' (left) and 'female' (right) brain, popularised in the 1980s book, *Men are from Mars, Women from Venus,* by John Gray. This overtly simplistic view of the relationship between the right and left hemispheres in the brain is loosely based on the fact that the left hemisphere specialises in language, maths and logic. It is analytical and is therefore used extensively in problem solving activities or when sequential processing is required, whereas the right hemisphere processes intuition, imagination, creativity and insight. The right brain processes things in a more holistic way and is more emotional, learning well from rhythm and music, images and pictures. While it is true that these areas in the brain ignite in different ways, the differences are more nuanced and complex than are claimed in the popular press.

It is accepted by neuroscientists that one half-brain is 'logical' and 'analytical' while the other is more 'intuitive' and 'creative'. Both halves play important roles in logical and intuitive thinking, as well as in analytical and creative thinking. All of the popular distinctions involve complex functions, which are accomplished by multiple processes, some of which may operate better in the left hemisphere and some of which may operate better in the right hemisphere. However, the overall functions cannot be said to be entirely the province of one or the other hemisphere.

For example, when a person is speaking, a region near the front of the left hemisphere controls the movements of the tongue, lips and vocal cords, but when a person is singing, the corresponding part of the right hemisphere performs these functions. Similarly, in order for a person to understand language fully they need to understand syntax that is the structure of sentences, which is better accomplished by the left hemisphere. They also need to understanding inflection and tone which is accomplished by the right hemisphere. Meaning is therefore deciphered only when both hemispheres are working together. In this sense, the two hemispheres work in symbiosis within a single system linked by a complex rope-like structure of nerve fibres known as the corpus callosum. The hemispheres use this bridge to communicate with each other and carry messages from one side of the brain to the other. The two halves work together and they are not isolated systems that compete or engage with one another in some kind of mental tug-of-war.

When it comes to learning, we can think of the left hemisphere as the activity attention network that helps us engage in externally directed tasks. It is focused on goal directed performance, engaging with the immediate concrete world outside our brain, which makes up much of our lives. It is where we plan for future actions, generate ideas, use tactics and consider what strategies worked before. On the other hand, we can think of the right hemisphere as the imagination network activity or the subjective realm of our 'inner experience'. This is our 'drifting mind', our 'wandering mind', our 'fantasy mind' or our 'creative mind'. The imagination network takes up half of our mental lives. It enables us to construct personal meaning from experiences, mental stimulation and the ability to acquire perspective of self and others (Dewey 1938).

The importance of both hemispheres in learning and healing is described with acute understanding by Jill Taylor, a neuroscientist, in her book, *My Stroke of Insight*. Her book describes how a stroke within the left side of her brain, left her unable to walk, talk, read or write. It took her nearly a decade, but she 'retaught' her brain and once again lives a normal life. However, she came to value her right brain's strengths: 'I shifted from the doing-consciousness of my left brain to the being-consciousness of my right brain ... I stopped thinking in language and shifted to taking new pictures of what was going on in the present moment' (Taylor 2008: 132).

The Four Lobes

In addition to the cerebral hemispheres, the cerebral cortex is divided into four main regions or lobes: the frontal lobe, located at the front of the head; the parietal lobe, located near the top of the head; the temporal lobe, located behind the temples of the head; and the occipital lobe, located at the back of the head.

These areas are distinguishable by ridges (known as gyri) and grooves (known as sulci).

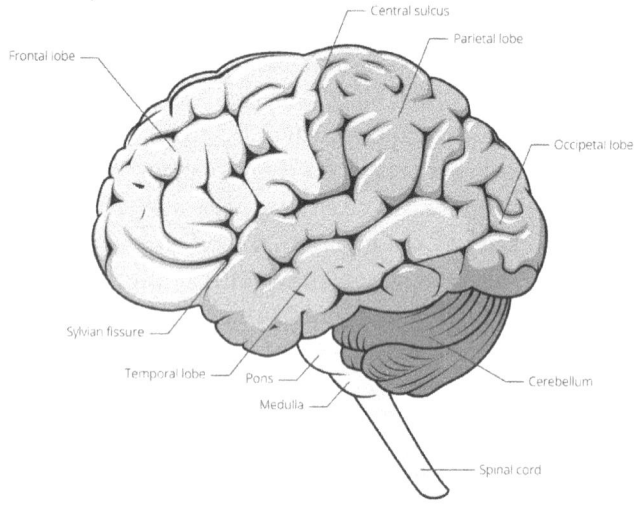

Figure 3: Human Brain Four Lobes – Image by *Oleksandr Pokusai*/<stock.adobe.com>

Brain scanning technology, such as MRI, allows neuroscientists to see beneath the hemispheres and learn more about which brain functions are associated with these different parts of the brain. Lying just behind the forehead, the frontal lobe is the command-and-control centre of the brain, monitoring its ability to control important cognitive skills such as planning, reasoning and judgement. The frontal lobe also contains what some might call the social being or personality, whereby decisions are made that are unique to each individual. Recall the story of Phineas Gage's accident and subsequent personality change. This area is vital to self-awareness, emotional self-regulation and resolution of conflict. Because most of the working memory is located here, it is the area primarily concerned with consciousness and focused attention. Attention may seem continuous, but maintaining focused attention can be difficult. The ability to filter out other possible objects of attention is thus crucial to studying and learning. Because the frontal lobe matures slowly, emotional regulation is not fully operational in adolescence, teenagers are more likely than adults to resort to high-risk behaviour.

Near the top of the head is the parietal lobe which interprets, integrates and coordinates sensory information with action. Between the parietal and frontal lobe is a band across the top of brain from ear to ear called the motor cortex. This strip controls body movement and, as we will see later, works with the cerebellum and basal ganglia to coordinate the learning of motor skills. It accounts for our body's awareness and tactile perception such as the ability to catch a ball by guiding our movements using real time information about its direction and speed.

> **Podcast**
>
> Michael Hobbiss was interviewed in 2018 by the *Learning Scientist* Podcast team about attention and the classroom. Hobbiss says that attention is captured and effected in two ways (a) through the object or stimulus that is capturing attention, and (b) through prior knowledge, interest, motivation, strategies used to perform the task. Both interact to influence attention. He says that we tend to think of attention as a resource that we have or don't have but this may not be a useful way to think about it. Many environmental factors influence our attention. This means that there is enormous potential to improve attention, particularly by removing distractions for example, putting your phone away, using pictures effectively, consciously choosing the type of music you listen to.
>
> Available from: <https://www.learningscientists.org/learning-scientists-podcast/2018/7/4/episode-22-attention-and-the-classroom-with-michael-hobbiss> (23 minutes)

Positioned above the ears is the temporal lobe which deals with language function and some parts of long-term memory. It encodes memories and works as a bridge between vision, hearing, language and facial recognition. While the primary motor cortex and related areas in the frontal lobe are responsible for the motor aspects of speech production, the temporal lobe contributes to the selection and retrieval of words and the grammatical structure of sentences, making it much more important to language comprehension.

At the back is the occipital lobe, the smallest of the lobes, which is used almost exclusively for visual processing and as the colour centre.

Finally, the cerebellum as seen in the above diagram is the most prominent rearmost part of the brain. It is sometimes called the 'little brain' because it resembles the wrinkled appearance of the crown of the skull, but its grooves and bulges are finer and organised into more regular patterns. It represents about 10 per cent of the brain's total volume but contains 50 per cent of its neurons, demonstrating the important role it plays in the control of voluntary movement, balance, posture and coordination. Because the cerebellum monitors impulses from nerve endings in the muscles, it is important in the learning, performance and timing of complex motor tasks. Receiving inputs from the eyes and ears, the cerebellum despatches instructions back through the brain stem to other regions in the brain. This part of the brain is like the conductor in an orchestra. It stores learned sequences of movement and coordinates and fine-tunes messages from elsewhere in the brain to create fluid body movements and habits, such as touch-typing or allowing a hand to bring a cup to the lips without spilling its contents.

Many studies have shown how the cerebellum is involved in learning sequences of action, linking those action sequences to form larger units. It is noticeable that, when musicians are practising a piece that they are learning, they don't correct individual notes if they make a mistake. Instead, they go back and repeat whole sequences. This emphasises how important the sequencing of actions is. It encourages the cerebellum to integrate whole chunks to create larger units of skilled performance. The same principle holds for language and reading and many other aspects of cognition. We know this because damage to certain parts of the cerebellum can eliminate the ability to perform precise movement or to adapt them to changing conditions (Hayes 2018).

The Inner Regions of the Brain

Together with the cerebellum, the brainstem is the bottom most region of the brain. It is the oldest section of the human brain which evolved about 500 million years ago (Striedter 2012). It is the site of the most basic survival automatic 'housekeeping' behaviours that keep us alive, such as monitoring breathing, controlling the digestive system, as well as regulating our heartbeat and managing our sleep-wake cycle. This is sometimes referred to as the reptilian brain. The brainstem also houses the Reticular Activating System

(RAS), responsible for the brain's alertness, which will be explained in more detail in the chapter on learning.

The upper part of the brainstem forms a bulge (called the pons) like a widening upright stalk that locks into the centre of the brain, and reaches up to the dome where the two hemispheres sit. In effect, it acts like a relay station between the lower and higher regions of the brain. This enables it to act as the means of communication between the brain and the rest of the body, connecting the spinal cord which stretches from the bottom of the skull to the tail bone.

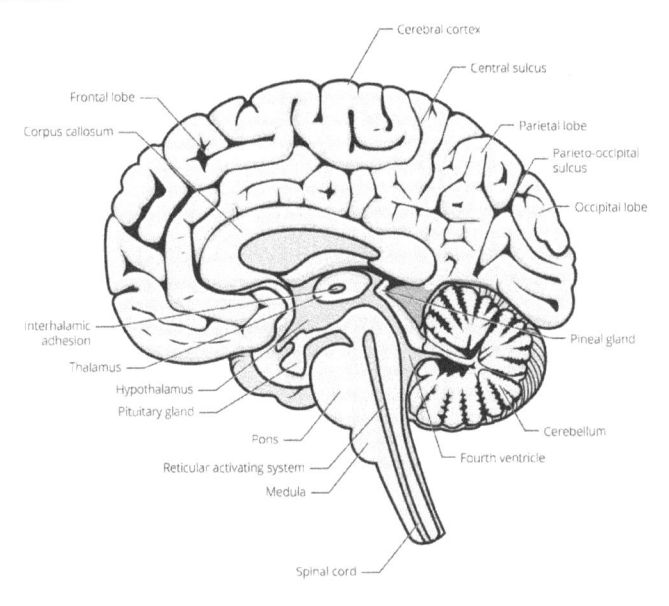

Figure 4: Saggital Section (Interior) – Image by *Oleksandr Pokusai*/<stock.adobe.com>

From the spinal cord, nerves reach out to every part of the body, regulating the central and peripheral nervous systems. Without the spinal cord, the brain would not be able to transmit information to the rest of the body or receive signals from all parts of the body.

The Thalamus and the Limbic System

Most learning begins when our sensory organs are stimulated through various experiences. Deep in the centre of the brain, there is a large area of packed cells above the brain stem and below the cerebrum known as the thalamus. This is separated into two halves, and it acts as a kind of relay station for sensory information and for motor signals going to the muscles. Some of these structures are vital in the coordination of fine movement and in the formation and maintenance of habits (Carter et al. 2019). Parkinson's disease is a powerful example of what happens when this area becomes damaged.

The egg-shaped thalamus is where all incoming sensory information, with the exception of smell, the most primitive sense, passes through before being sent upstairs, so to speak, to the cortex. From there, sensory information is directed to other parts of the brain for additional processing. The thalamus is the gateway to the neocortex. More than a simple relay system, the thalamus appears to decide whether the various inputs merit being sent to the cortex for conscious consideration. Many neuroscientists suspect that the thalamus plays a critical role in human consciousness, as it is the region best placed to receive, combine and analyse information because of its capacity in making fast links between multiple brain areas.

Nestled above and around the brain stem lies a complex set of interconnected structures of different shapes and sizes known as the limbic system. The limbic system is responsible for regulating emotions, forming memories and shaping behaviour. In recent years this structure in the brain has generated much interest because it is connected to emotions. According to Damasio, 'emotion was largely neglected by neuroscience during most of the twentieth century, but is now the focus of intense scrutiny, and not a moment too soon considering its importance in human lives. The neurobiological underpinning of the emotions has begun to be elucidated and it has become clear that the brain handles emotions with the help of different components'

(2000: 14). These component structures include what is now referred to as the limbic system.

Most of the structures in the limbic system are duplicated in each hemisphere of the brain. The term limbic is derived from the Latin *limbus*, meaning 'border' or 'ring' and it has a wraparound shape with nerve fibres connecting lower regions and higher regions of the brain. It is situated approximately in the centre of the brain and its placement between the brainstem and the cerebrum facilitates the interplay between emotional processing and rational thinking. The limbic system is involved in instinctive survival behaviours including the generation of emotions, and for this reason is often referred to as our emotional brain. We can say that one of the main functions of the limbic system is to analyse our experiences and make them meaningful.

The limbic system is a varied collection of structures surrounded by an area of the cortex referred to as the limbic lobe. Visually, the lobe structures resemble the head of some beast with formidable, swept-back horns. It circles the thalamus and has a tail-like arch which sweeps forward, and then backwards, to form an almost complete circle. Nerve fibres link all these parts intimately and also connect them to other areas of the brain. In reality, it is not a unitary system, as different parts carry out similar functions in a coordinated circuitry manner.

What part of the brain actually decides when, where and how we move? The answer lies buried deep within the limbic system. Basal ganglia is a group of important nuclei located deep in the centre of the brain which controls many aspects of voluntary movement, as well as some aspects of cognitive functioning and emotion (Parent, 2012). Through a complex system of circuits, the basal ganglia receives inputs from the cerebral cortex and send outputs to the thalamus and brainstem to impact motor and cognitive functions (Redgrave et al. 1999).

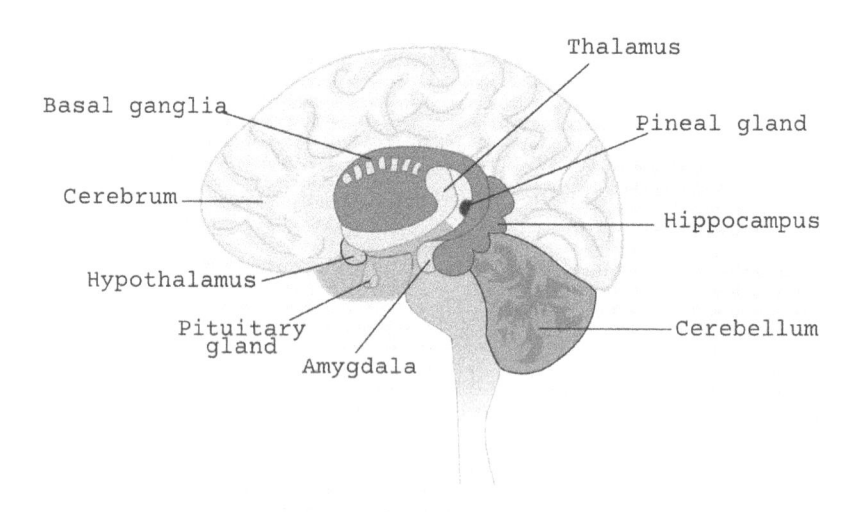

Figure 5: The Limbic System – Image by *LuckySoul*/<stock.adobe.com>

The formation of the limbic system reaches back deep into our evolutionary past. It is involved in the regulation of motivated behaviour, including the five *Fs* of motivation: fleeing, feeding, foraging, fighting and sex behaviours (an old physiological joke). Foraging, in this context, means learning – what we do when we search online for one item, but end up an hour later having moved from site to site. Three structures, in particular, are important in learning and memory. They are the hypothalamus, hippocampus and amygdala. We will briefly introduce them here as understanding these connections will be discussed in later chapters.

The Hypothalamus

The hypothalamus, one of the most ancient parts of the brain, is responsible for the release of chemical signals into the bloodstream. The word 'hypo' means below and the hypothalamus sits directly below the thalamus. This allows it to send messages about the emotional state of our brain out into the body. It thus acts as a processing centre for many other regions of the brain and the body, confirming that learning is not all in our head (Hannaford 2007).

The hypothalamus is essentially involved in the learning process in two different ways. Firstly, millions of neurons carry signals back and forth. This is the way the brain senses what is happening moment by moment in and around the rest of the body and the way it sends out instructions for what the body should do. Secondly, the brain produces hormones that are sent into the bloodstream, and these hormones are specific messages about what the brain is experiencing. Likewise various parts of the rest of the body produce other types of hormones that are picked up by the brain and tell it what is happening throughout the body.

In short, hormones react to the firing of neurons but they are slower and can be less precise. However, it is the hormones that are responsible for our feelings, which colour our experiences. Stress and learning are a good example of how this works. A stressor can be defined as anything from the outside world that disrupts our homeostatic balance (our comfortable condition in the body). The stress response is anything we do to restore this. Our sympathetic nervous system mobilises us to deal with threats, real or imagined.

For instance, we may remember a time when we were a little anxious, tense and unable to pay attention when required to give an oral presentation to our class as part of a graded assignment. Not surprisingly, this required us to move out of our comfort zone and into our stretch zone for a short period of time. This is called 'good' stress as some stress in learning is beneficial. In preparing and practising the presentation, some stress helps our learning by producing more oxygen to the brain, so neurons have more energy resource at their disposal. In this stage of engagement, we learn better. Also, our adrenal glands release the hormone adrenaline that increases breathing, heart rate and blood pressure. Adrenaline causes a rapid release of glucose and fatty acids into our blood stream to give us energy. Our senses become keener, our concentration sharper and afterwards the feelings that were part of the experience seem to help us remember it as positive learning.

In most learning situations, moderate or manageable stress in the form of challenge improves performance. However, at a point determined by the individual's capacity to accommodate stress learning performance dips dramatically. Higher levels of anxiety characterise this phase of learning. In what ways might stress get in the way of learning? This will vary for each individual but lifestyle factors, as well as internal perceptions, play a significant role. Whatever triggers a prolonged period of stress, the adrenal gland also releases another chemical into the blood, cortisol. Once in the brain, cortisol remains much longer than adrenaline and can affect the manner in which neurons function. Too much cortisol adversely affects learning in two ways. First, because it diverts glucose to other parts of the body, the amount of energy that reaches the hippocampus is diminished. This adversely affects cognitive functions like concentration and creativity. Second, it interferes with the functioning of neuronal synaptic activity and neurotransmitters. Therefore, an increase in cortisol leads to a reduction in the retention of new information. The practical result can manifest as difficulty in thinking or accessing some elements of previous learned material.

The Hippocampus

The hippocampus is one of the key structures in the brain. It plays a major role consolidating learning and in converting information from short-term memory via electrical signals to long-term memory. It also plays a significant role in how we understand and navigate our surroundings. It is a structure that is particularly important in forming new memories and in connecting emotions and senses, such as smell and sound, to such memories. It is also responsible for turning experience into new neural pathways that are stored for future reference. This process is essential for the creation of meaning (Barrett 2018). The hippocampus acts as a memory indexer by sending memories out to the appropriate parts of the cerebral hemisphere for long-term storage and retrieving them when necessary. In

Alzheimer's disease, the hippocampus is one of the first regions to suffer damage.

The Amygdala

Attached to the end of the hippocampus is the amygdala, a tiny almond-shaped structure. It is the area most active when we feel emotions. In his book, *The Emotional Brain*, LeDoux (1999) describes how sensory signals go directly to the amygdala, before we are even aware of them. As such, the amygdala is constantly monitoring our experience to see how things are. Our amygdala is like a danger warning signal that says, 'this is bad for me'. It helps decide meaning but it does not solve problems, create new ideas or plan new actions.

The amygdala functions as a memory or association system for events that are emotionally significant such as fear and desire. However, the amygdala does not produce conscious memories; rather, it produces unconscious responses to stimuli. These responses are felt rather than thought, such as when you have an uncomfortable intuition or feeling about a person or situation but cannot explain why. It is interesting to realise that the two structures in the brain mainly responsible for learning are located in the *emotional* system. This insight has important relevance to the theory and practice of educational neuroscience, as will be explored in Section 3 of this book.

The Integrated Brain

The two brain hemispheres, the four lobes, the thalamus and the associated parts of the limbic system including the hypothalamus, the hippocampus and the amygdala are all a complex set of structures, with a key role in the brain's processing of thought, motivation and emotion. However, the human brain is very different from a computer. Whereas a computer works in a linear, step by step fashion – albeit very fast – the brain works associatively, as well as linearly, to carry on thousands of different processes at the same time, comparing, integrating and synthesising as it goes (Russell 1980). Thus a person often finds that in conversation the brain is not behaving linearly but racing on in different directions, exploring new ideas and the ramifications of what is being said, and at the same time taking in subtle changes in intonation, body language and facial expression, as well as the linear sequence of words.

According to Amen, what happens in a normal functioning brain is as follows: 'information from the world enters the brain through your limbic system, where it tags the information as meaningful, safe or dangerous, then it travels to the back part of the brain (temporal, parietal, and occipital lobes), where it is initially processed and compared with past experience, and then to the front part of the brain for you to consider and then act on it. The transmission of information from the outside world to your conscious awareness in the front part of your brain happens almost instantaneously' (2011: 5–6).

We would caution against the natural tendency to see the parts of the brain as separate, rather than the systemic whole to which they belong. Increasingly we realise that no one part of the brain can be responsible for all that we experience. We are, if nothing else, designed to respond holistically to the environment in which we survive and thrive. Like that of a supremely engineered aircraft, such as the Airbus 380, all its integrated systems are designed to work together to accomplish a specific goal. No one part alone would accomplish much as the power of the aircraft lies in how these parts all work together. In the case of the brain its component parts interact in a similarly complex fashion.

Conclusion

This chapter provides a basic foundation for understanding the parts, functions and interconnections within the brain. To better understand learning, it is essential that we study individual and holistic perspectives on brain structure and function, to help learners and educators to tailor their approaches to learning to meet varied needs. When opportunities are created that appreciate the interconnected nature of the brain, there is great potential to deepen learning, as we will demonstrate in later chapters.

References

Aldersey-Williams, H. (2014). *Anatomies: The Human Body, Its Parts and the Stories they Tell.* London: Penguin Books.
Amen, D. G. (2011). *Change Your Brain, Change Your Body.* London: Piatkus.
Calvin, W. (1997). *How Brains Think: Evolving Intelligence, Then and Now.* London: Weidenfeld and Nicolson.
Carter, R., Aldridge, S. and Page, M. (2019). *Brain Book.* London: Dorling Kindersley Limited (DK). Random House.
Clancy, J. (2018). *The Secret Life of the Human Body.* London: Octopus Publishing Group.
Damasio, A. (2008). *'Descartes' Error: Emotion, Reason and the Human Brain.* New York: Random House.
Dewey, J. (1938). *Experience and Education.* New York: Macmillan.
Feldman Barrett, L. (2020). *Seven and a Half Lessons about the Brain.* London: Picador.
Gilbert, I. (2008). Foreword. In A. Curran (Ed.), *The Little Book of Big Stuff about the Brain.* Carmarten, Wales: Crown House Publishing.
Gray, J. (1992). *Men Are from Mars, Women from Venus.* New York: Harper Collins Publishers.
Hannaford, C. (2007). *Smart Moves: Why Learning Is Not All in Your Head* (2nd edn). Salt Lake City, UT: Great River Books.
Hayes, N. (2018). *Your Brain and You: A Simple Guide to Neuropsychology.* London: John Murray Learning.
LeDoux, J. E. (1999). *The Emotional Brain: The Mysterious Underpinnings of Emotional Life.* New York: Simon & Schuster.
MacLean, P. (1990). *The Triune Brain in Evolution: Role in Paleocerebral Functions.* New York, NY: Springer.

Parent, A. (2012). The History of the Basal Ganglia: The Contribution of Karl Friedrich Burdach. *NM* 03, 374–379.

Redgrave, P., Prescott, T. and Gurney, K. N. (1999). The Basal Ganglia: A Vertebrate Solution to the Selection Problem? *Neuroscience* 89, 1009–1023.

Schultz, W. (2006). Behavioral Theories and the Neurophysiology of Reward. *Annual Review Psychology* 57, 87–115.

Striedter, G. (2012). *Principles of Brain Evolution.* New York: Random House.

Taylor, J. (2008). *My Stroke of Insight.* New York: Penguin Books.

CHAPTER 3

The Importance of Neurons

'All that we know, all that we are, comes from the way our neurons are connected'.

(Berners-Lee 1999)

What is it about the human brain that allows it to generate such complexity in areas such as language, reason, memory, emotion and intuition? The main architect of everything that happens in the brain is a very special nerve cell called the neuron. Through a combination of chemicals and electricity, neurons communicate with each other and pass messages around from one area of the brain to another. The primary task of the brain is to help maintain the whole body in an alert state relative to the environment, in order to maximise its chances of survival. The brain does this by registering nerve signals and in the process generates actions which in turn create subjective experience.

The nervous system is made up of two regions: the central and the peripheral. The central nervous system is the brain's main control centre, responsible for co-ordinating all the processes and movements in the body. The brain and spinal cord constitute this central nervous system and it is connected to the peripheral nervous system by a collection of 43 different pairs of nerves. Thousands of peripheral nerves stretch out through all parts of the body, carrying tiny electrical signals to and from the central nervous system. It is an information dual motorway, buzzing with activity as it channels all the signals sent to and from the brain. See figure 6 for an illustration of the central nervous system.

The one structural characteristic which distinguishes neurons from all other types of body cells is that neurons are specialists in communication. They process, relay and store information. Their unique structure and function allow them to transmit signals to and from the brain and spinal

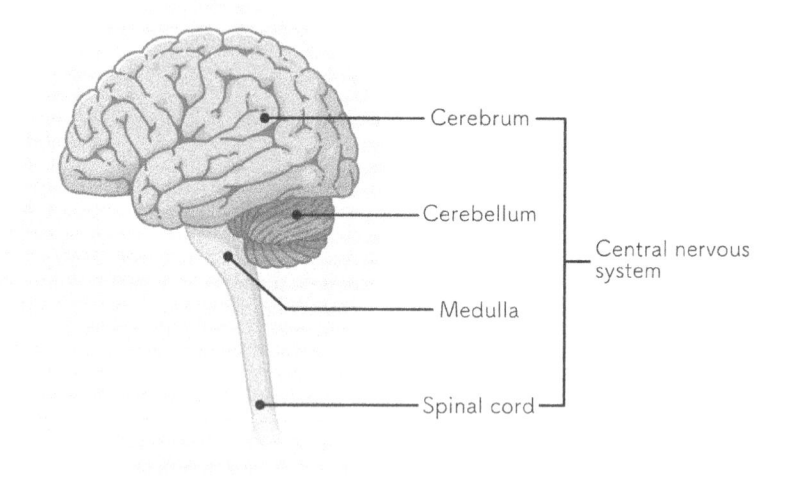

Figure 6: Central Nervous System – Image by 川本まる/<stock.adobe.com>

cord. The job of the neuron is to communicate messages to other neurons (Swaab 2014). Complex decisions about behaviour are made within the brain by complicated circuits of neurons. It is these interconnections that are critical to information processing and learning.

The key to this relatively small organ, weighing an average of three pounds (1.4 kg), effortlessly accomplishing a myriad of tasks, is in its rich diversity of neurons and the way they send messages between each other to form pathways and associated connections of networks. There are billions of neurons in the brain – approximately 100 billion in a mature brain (Feldman Barrett 2021). Each individual neuron makes contact with about ten thousand other neurons, so that the actual number of connections between the neurons in the brain is in the trillions. And yet, only a small proportion of the brain cells (about 10 per cent) are neurons.

Neurons need a supporting structure made up of other cells called glia. The early anatomists thought of the glia as a 'glue', hence the name. Some glia cells perform the function of protecting the neuron, while others are used in the healing process of damaged neurons. Lastly, others (especially

Schwann cells) form a substance called myelin, a fatty sheath wrapped around the neuron that acts as a kind of an electrical insulator. Myelin plays a pivotal role in the consolidation of learning – it is why people can still recall the multiplication tables they first learned in school, or the words of a song from their late teens.

From the perspective of educational neuroscience, the brain can be studied at two levels. The lower level concerns the workings of individual neurons and their interconnections, establishing what makes them fire and how the electrical pulses relay signals between them. At a higher level, the brain can be regarded as a fantastically complex network around which electrical and chemical patterns interact. If higher mental processes are associated with patterns of neural activity, then it is the neural networks that are most likely to illuminate human consciousness and the learning process. The complexity of how learning occurs is embodied in a doubly complex tangle of neurons in our brains – the local complexity of thousands of units within each network and the overall complexity of the feedback loops between the networks.

As we will see, learning results in physical changes in the brain, and these changes are unique for each individual. In order to understand the complexity of our mental processes, we must therefore have a basic understanding of the workings of the neurons and associated networks in the brain. This chapter will explore the important connections between brain cells and neurons that enable us to store all our memories, our knowledge of the world, our hopes and fears etc. It is these connections that underpin all learning, whether from formal schooling or from our life experiences.

The Structure of a Neuron

Neurons are specialised to produce electrical signals and communicate with each other. All sensations, movements, thoughts, memories and feelings are the result of signals that pass between them. What distinguishes the neurons from the billions of other cells in the human body is their shape. Most other cells are neat and compact, but neurons are not as tidy.

They also come in many different shapes and sizes, stretching, twisting and dividing in hugely complicated neural networks. The human brain contains at least several hundred different types of neurons. The three most typical types are the unipolar, bipolar and multipolar.

A schematic drawing of a single neuron is shown in Figure 7. At its simplest, each neuron looks like a little tree, with bushy branches at the top, a long trunk and roots at the bottom. The bushy branches, receive signals from other neurons, and the trunk, sends signals to other neurons, through its roots. Despite the fact that neurons come in many different sizes and shapes, they all share a common design feature which can be subdivided into three parts: the cell body, the dendrites and the axon. Each of these parts has a distinct function.

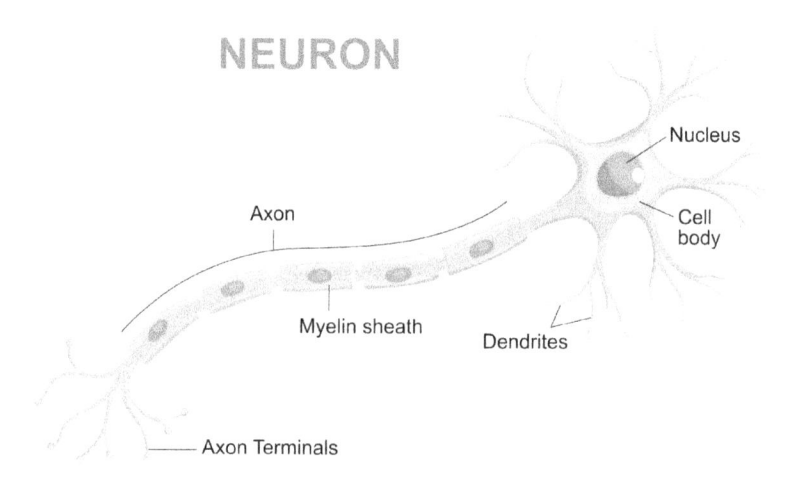

Figure 7: The Structure of a Neuron – Image by *L. Darin*/<stock.adobe.com>

The round body of the neuron is known as the cell body or soma. It serves all the necessary functions needed to maintain the health and integrity of the neuron, such as manufacturing energy from appropriate chemicals that it takes in from the blood stream, as well as protein production. In addition, the outer covering of the cell body collects electric charge which also contains a combination of genes (DNA) that gives it its unique characteristics. An individual cell body extends out in two directions.

The dendrites are thin filaments, forming a winter leafless tree-like structure. The Greek word *dendron* means tree, and the dendrites of a neuron are so called because of their branch-like appearance. These slender branches are the neuron's 'input' compartment – they receive electrical signals from the neighbouring neurons, before conveying them to the cell body. Each of these dendrites can conduct a weak electrical charge along its membrane to the body of the cell. If enough of these weak charges arrive at the cell body within a small interval so that its membrane becomes charged to a critical point, they will then discharge its electricity along the long filament leading out of the cell, which is known as the axon.

Each neuron has only one axon which is the 'output' department. It is usually much longer and thicker than the dendrites. Importantly, at the end of the axon are branching filaments which form bushes similar to those on the dendrite side. At the end of each axonal branch, there is a small lump, called a synaptic knob, which makes a connection with other neurons. The important difference is that the axonal branches conduct electricity outwards from the cell body, while the dendrites conduct it inwards.

As the human brain matures, a process of myelination of axons occurs within the nervous system. As connections between neurons become more frequent, a myelin sheath forms around the axons to enable and ensure efficiency of communication between one neuron and another. We now know that axons with myelination conduct impulses (messages) much faster than axons which are unmyelinated (Waxman 1980). According to Williamson and Lyons, an increasing body of evidence suggests that myelin may play a role in circuit plasticity and that myelin 'may be an adaptable structure which could be altered to regulate experience and learning' (2018: 1).

How Does One Neuron Communicate with Another?

The starting point for neuron communication is the electrical signal which any one neuron fires in response to a stimulus. For example, when a person performs a movement, or entertains a thought, a neuron becomes active; it is said to fire, spike or generate a nerve impulse. Whenever a neuron is stimulated, an electrical signal travels down its axon, like a line of falling dominoes, to the axon terminal. This firing is called the action potential.

When scientists first realised that electricity is conducted within nerve cells, they believed that these cells also sent messages to one another by means of electrical currents. This has turned out to be true only of very few types of neurons. In general, it has been found that there is a gap, known as a synapses or synaptic junction, between one neuron and the next. Although an axonal branch leading out of neuron (A) may come very close to a dendrite leading into neuron (B), the two do not touch and no electricity is conducted from one to the other. How then does a message get across the gap? By means of a chemical known as a neurotransmitter.

Put it another way. Electrical impulses in the brain are like runners in a 4 × 400 relay – they cover long distances at great speed (Scott and Grice 1998: 107). Neurotransmitters are like the batons exchanged between the runners. Our brains exchange 'batons' when transmitting the signals needed to process information, regulate emotion, learn and store experiences and keep us alive. Some neurotransmitters are inhibitory, some are excitatory and in this way they make us think, feel and act.

There are several types of chemicals that function as neurotransmitters, but the difference between them is not important for our present purpose and will be covered in later chapters. In short, these neurotransmitters are stored in little packages or vesicles near the tips of the axonal branches. When the electrical current – in the form of the action potential – reaches the tips of the branches it causes the neurotransmitters to be released into the fluid-filled gap. The neurotransmitter molecules then float across the synapses, where they meet the dendrites of the next neuron, and bind themselves to special proteins called receptors, much like a boat docking at port.

A receptor, however, will not let some kinds neurotransmitter fit into it – it must be a specific match whereby the molecules correspond perfectly. Once a neurotransmitter is locked into the receptor, and bound to it, the creation of a new chemical acts as a trigger for the next series of events to unfold. If the neurotransmitter has the right shape, it will fit the receptor on the next neuron. Each neuron releases its own neurotransmitters, stimulating the next cell to do likewise, and so on. In this way, millions of messages are passed from neuron to neuron every minute of the day at astonishing speeds (Magrini 2019).

Although the process always occurs in the same way, its individual stages contain aspects that can vary. For example, the number of neurotransmitters emitted by a neuron can be greater or smaller. Moreover, the neurotransmitter must be taken back into the sending neuron shortly after it is emitted, as otherwise the absorbing neuron would continuously be activated, and not just when a message needs to be sent. This process of re-uptake can be quicker or slower – the slower it is, the more strongly the second neuron is activated. In addition, the second neuron can have more or fewer receptors per unit on its dendrite at the point of the synapse. The more receptors, the more strongly the receiving neuron is activated. All these variables affect the strength of the synapses between the neurons.

What this tells us is that with stimulation and the passage of time, neurons grow more axons and dendrites which in turn form complex neural pathways. Once laid down, these pathways ensure rapid communication between areas of the brain that may be relatively very far apart. In fact, dendrites are the core of what determines the computational power of a single neuron. In this way, wider pathways, like our habits or interests, begin to develop. If pathways are not regularly used, they gradually fade away. Neuroscientists have labelled this process *pruning*. This refers to the brain's ability to change and adapt throughout life. This is not merely a superficial change but, rather, a fundamental transformation in the physical structure of the brain. Synaptic connections can become stronger or weaker depending, for example, on what we pay attention to and by our habitual reward system. It is this wiring of neurons that makes each of us the individuals we are.

How Neural Networks Are Formed

Neurons create a network through the formation of synaptic connections. These connections allow neurons to communicate and interact with each other to process information. Interconnected networks can collectively process information thus serving as the basis for all our perceptions, thinking, memory, and the planning of our actions (Mukherjee 2022). The connections between neurons are not confined to those within individual networks. In order for the brain to process information gained from perception and to plan action to further our goals through the use of this information, the various networks must be connected with one another.

Feldman Barrett (2021) suggests the metaphor of the global air-travel system to describe these complex neural networks. The air-travel system is a network of about seventeen thousand airports around the world. Whereas the brain carries electrical and chemical signals, this network carries passengers. Each airport runs direct flights to some other airports, but not to every other airport. Instead, some airports take the burden off the rest by serving as hubs. Our brain network is organised in much the same way. Its neurons are grouped into clusters like airports. Most of the connections in and out of a cluster are local so like an airport, the cluster serves mostly local traffic. In addition, some clusters serve as hubs for communication. Brain hubs, like airport hubs, make a complicated system efficient. They allow most neurons to participate globally even as they focus more locally. Thus, such networks or hubs form the backbone of communication throughout the brain.

The Importance of Neurons

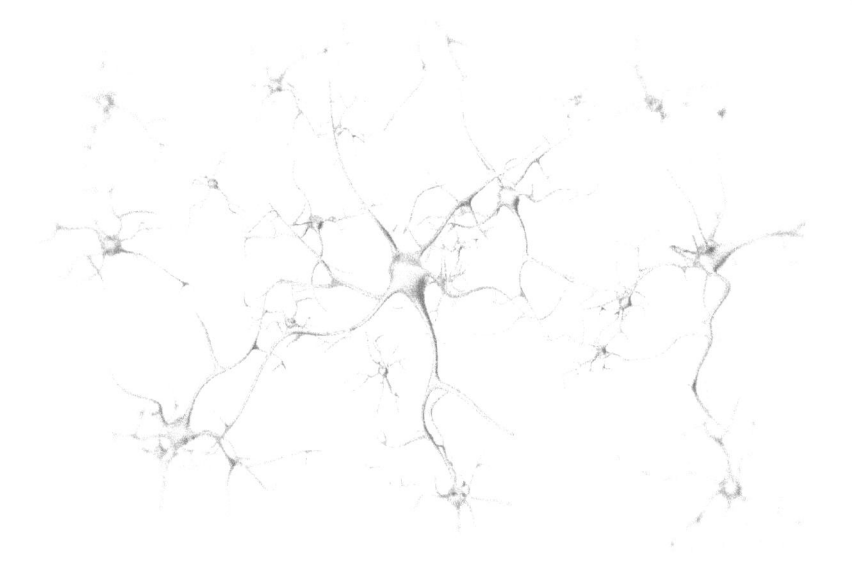

Figure 8: A Neural Network – Image by *rost9*/<stock.adobe.com>

Networks are absolutely essential structures. To continue with the analogy, when a major airport hub like London's Heathrow goes down, flight delays or cancellations ripple across the world. So imagine what happens when a brain network goes down. The resultant damage may cause learning difficulties, depression, schizophrenia, dementia, and other degenerative disorders such as Alzheimer's or Parkinson's disease. Networks are points of vulnerability because they are points of key efficiency. Scientists believe that, over evolutionary time, through natural selection, neurons organised into this kind of network format because it is powerful and fast, yet energy-efficient and still small enough to fit inside the skull. Clearly, our brain network is never static, as the connections between the neurons keep changing in never-ending patterns.

The Connections between the Neurons Keep Changing

One of the most important features of the connections between neurons is that they can be modified as a result of experience. This property of the synapses is called *plasticity*, as a soft plastic object can easily be shaped into different forms. So, what causes the synapses to change and how does this change take place?

Changes in the synapses occur according to the principle known as Hebb's rule: every time a synapse fires after receiving an excitatory input from another neuron, the synapses linking the two neurons is strengthened. From the outset, all the neurons in a network are connected with one another, whether directly or at one or two removes. However, new synapses are rarely formed after the early period of neural development, as nearly all learning in the brain occurs through the subsequent strengthening or weakening of the links between the synapses.

Back in the 1950s, long before brain scanning had become a reality, Hebb proposed that learning happens because cells form cell assemblies. The more often one neuron stimulated another, the stronger its connection would become. Hebb believed that the synaptic knob making the connection between the cells would grow larger so that their effect was stronger. However, modern research shows that he was very nearly right, as the synaptic knob itself does not grow but the number of vesicles and receptors associated with that synapse increases the more the connection is made. This is known as synaptic plasticity, and it means that the next nerve cell gets a stronger message when the first one releases its neurotransmitters, making it more likely to fire. This is how we develop learning any new academic skill such as writing an essay; effectively, it is by training groups of neurons to become linked to one another through practice which will eventually result in reinforcement.

Why does this happen? It is all to do with how adaptable we are as a species. As we know, human societies are found in all sorts of different environments, from the frozen Arctic to baking deserts, from jungles to high mountains, from small towns and cities with over 10 million people or even in a small group in a space station, many miles from Earth. We can survive in all these environments, even though the survival skills needed are very different. But we do not survive by merely adapting physically – we also do it by learning (Winston 2010).

Learning and Neural Plasticity

Humans have an incredible capacity for learning and that is what the human brain is all about. Learning happens as the neuron cells in the brain strengthen particular synapses, weaken others and channel nerve impulses through unfamiliar pathways to produce a given effect. Our brain cells accordingly respond to the demands of our learning in two ways. One way is by building up the synapses – growing up the capacity of the synaptic knob and its associated receptor site, so that more neurotransmitters are released and picked up, making for a stronger message. The other way is through myelination. A neuron that fires only occasionally does not generally develop a myelin sheath. However, if a group of cells continue to be stimulated, as they will when you are learning something new, glial cells begin to wrap themselves around their axons. As we have mentioned, myelination helps the message to pass along the neurons more quickly, so we find the process of learning easier.

Much of the brain's computational power comes from this enormous possibility of interconnection, and much of it comes from the fact that brains are parallel processing machines, rather than serial processors. A lot of our learning happens when we are infants, but we continue to learn throughout our lives. This means that we are constantly placing demands on our brain cells, encouraging them to form new connections. The ability of brain cells to adapt is known as neural plasticity. It used to be thought that the brain would adapt only up to early adulthood and that after that the function of our brain cells was largely fixed. This idea of a static brain we now know is no longer the case. Yes children do recover from brain damage more easily from adults, but adults are also able to recover from many types of damage to the brain, by rechannelling neural impulses so that they form new pathways (Doidge 2007). We know, too that neurons can continue to grow and develop throughout our lives, as long as they receive the cognitive or physical demands needed to stimulate that growth.

Conclusion

When we think about something as complicated as the human brain, we cannot expect simple answers or fully comprehend sophisticated mental processes such as awareness, consciousness, imagination, memory and problem solving. In order to hold a single thought for just a fraction of a second, millions of neurons have to simultaneously connect. Most scientists agree that the way to understand the brain lies in understanding its circuitry but this has proved exceptionally difficult to study (Tornero and Soriano 2020; Winston 2010). The paramount problem is that the thin protrusions that connect neurons to each other are notoriously difficult to track, and there is an unimaginable number of them. Science continue generating so many insights on so many fronts. Take for example the sequencing of the human genome (DNA) which is not only revolutionising medicine but also revealing new perspectives on brain health and human learning.

Indeed in the past sixty years the human brain has been transformed from an inscrutable 'black box' as we uncover some of the brain's more tightly kept secrets. Perhaps another of the most significant insights from neuroscience research is the fact that the brain has the capacity to change throughout the lifespan. This, together with the idea of the brain as a parallel processor, and as an organ that can generate increasingly stronger networks, is currently driving an international team of scientists to create the most detailed map of the brain, identifying the locations and functions of its billions of neurons. A study known as the Brain Research through Advanced Innovative Neuro-technologies (BRAIN) Initiative involves laboratories across many countries and disciplines working to map the cells of rats and primates (including humans), in the hope of making critical discoveries about disease and brain function. This initiative is leading the charge in the development of new tools and understandings of the brain, much like early work on MRI and EEG scans before it.

References

Bernes-Lee, T. (1999). *Weaving the Web, The Original Design and Ultimate Destiny of the Worldwide Web*. San Francisco: Harper Business.
Doidge, N. (2007). *The Brain That Changes Itself: Stories of Personal Triumph from the Frontiers of Brain Science*. London: Penguin Books.
Feldman Barrett, L. (2021). *Seven and a Half Lessons about the Brain*. London: Picador.
Fornero, D. and Soriano, J. (2020). *Neuronal Cultures to Study the Brain and Neurological Disorders*. Health and Medicine, Research Outreach. https://researchoutreach.org/articles/neuronal-cultures-study-brain-neurological-disorders/
Gibb, B. J. (2012). *The Rough Guide to the Brain: Get to Know Your Grey Matter*. London: Penguin Group.
Jakel, S. and Dimou, L. (2017). Glial Cells and Their Function in the Adult Brain. *Frontiers in Cellular Neuroscience* 11 (24), 1–17.
Magrini, M. (2019). *The Brain: A User's Manual: A Simple Guide to the World's Most Complex Machine*. London: Short Books.
Mukherjee, S. (2022). *The Song of the Cell: An Exploration of Medicine and the New Human*. London: Penguin Random House.
Scott, T. and Grice, T. (1998). *The Great Brain Robbery: What Everyone Should Know about Teenagers and Drugs*. Wellington, New Zealand: Aurum Press.
Swaab, D. (2014). *We Are Our Brains: From the Womb to Alzheimer's*. London: Penguin Books.
The Brain Research Throught Advancing Innovative Neuro-technologies (BRAIN) initiative (2024). National Institutes of Health <https://braininitiative.nih.gov/>
Waxman, S. G. (1980). Determinants of Conduction Velocity in Myelinated Nerve Fibers. *Muscle Nerve* 3, 141–150.
Williamson, J. and Lyons, D. (2018). Myelin Dynamics Throughout Life: An Ever-Changing Landscape? *Frontiers in Cellular Neuroscience*, 12(1), 1–8.
Winston, R. (2010). *What Goes on in My Head? How Your Brain Works and Why You Do What You Do*. London: Dorling Kindersley.

CHAPTER 4

The Power of Plasticity

> 'Any man could, if he were so inclined, be the sculptor of his own brain'.
>
> (Santiago Ramon y Cajal 1897: xv)

For thousands of years, people believed that the sun went around the Earth. So widespread and firm was this belief that it was taken to be reality. In the sixteenth century, however, Copernicus put forward the radically different idea that the Earth went around the sun. His theory was not readily accepted and considered by many as heresy. It took a century of persuasion before the old reality was scrapped and a new reality adopted. Scientists refer to this process as the creation of a new paradigm, a term coined by philosopher and science historian, Thomas Kuhn, in his book, *The Structure of Scientific Revolutions*. A paradigm is like a 'super theory' providing the basic model of reality within a particular science. In short, a paradigm is a shared consensus that embodies the beliefs and values in a field of study that informs how problems are solved (Kuhn 1996).

Once accepted, paradigms are seldom questioned and usually govern the way a scientist thinks, and the way in which experimental observations are interpreted. To overturn 'accepted' science often requires decades of mounting evidence to demonstrate that the old theories no longer hold sway. The practice of medicine, to take one scientific endeavour, is replete with examples of conventional practice becoming obsolete, and newer theories initially ignored because they do not fit the old paradigm. This usually results in what Kuhn referred to as a 'paradigm shift', a change in a world view.

The paradigm shift in relation to brain science has predominantly occurred during the last 500 years. It was once thought that we were born with a given number of neurons, then lost a certain number every day as part of the wear and tear of life. Some organs of the body were acknowledged to have powerful regenerative abilities – consider for example how a bone will mend after being broken in several places, or how the liver recovers from the effects of excessive alcohol consumption. However, the brain was believed to possess very limited ability to heal itself after injury. One of the most profound intellectual shifts in contemporary neuroscience is the realisation that the brain has enormous capacity for self-healing and regeneration.

The changes that occur in the brain are often referred to as neuroplasticity. Studies using MRI brain scanning technology have shown that new neural cells and neural pathways are generated throughout one's life. This ability of the brain to reorganise itself by forming new neural connections is known as *neuroplasticity*. We now know that the way the neurons are wired together can be dramatically changed as a result of experience, learning and behaviour (Craver 2007).

Neuroplasticity: How to Rewire Your Brain

Plasticity is the term neuroscientists use to describe the lifelong ability of the brain to reshape, change and restructure itself based on new experiences. No other organ in the human body takes as long to develop or goes through as much change as the brain. If anything, the way our brain is wired might be as unique as our fingerprints.

The revolutionary idea of brain plasticity has a long history going back to 1890, when William James, the founder of experimental psychology in the United States, wrote that the brain is 'endowed with a very extraordinary degree of plasticity'. Because he did not have hard evidence to support his claim, the idea was not taken up. A few years later, in 1923, the Spanish anatomist, Ramon y Cajal discovered the way neurons communicate. In 1932, Charles Sherrington et al., received the Nobel Prize for discovering the function of neurons, and fifty years later Roger Sperry, of the California Institute of Technology, shared the 1981 Nobel Prize in

Physiology for his work on showing that there are important differences in function between the left and right halves of the brain (O'Shea 2005).

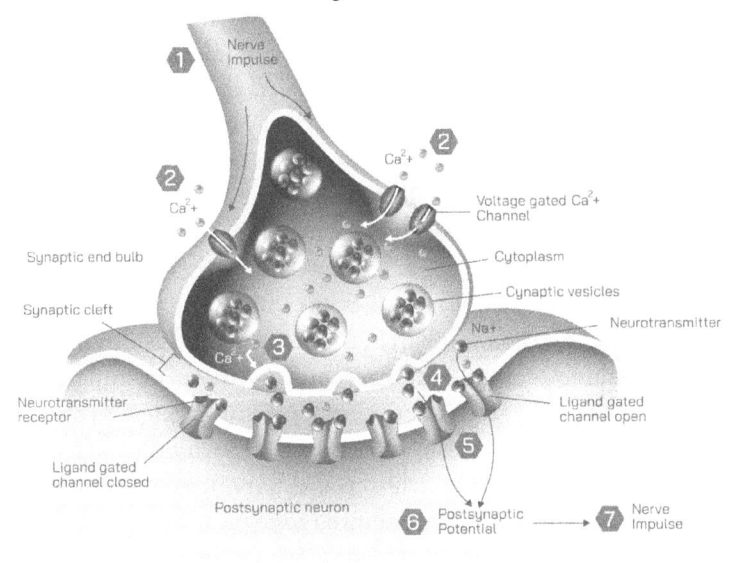

Figure 9: Neuroplasticity: Synaptic Density (png by Bokkyu Kim is licensed under CC BY-SA 3.0)

However, a conceptual framework to understand the idea of neuroplasticity in humans first emerged when the Canadian psychologist, Donald Hebb, published his classic book *The Organisation of Behaviour*. Hebb (1949) was curious about how learning and memory take place and realised that some structural change had to happen between neurons in order for learning to occur. If we learn a new skill some change must take place in the brain. He thought that if a group of neurons were repeatedly stimulated at the same time, then an active circuit, or what he called a 'cell assembly', might develop. If this circuit gets fired up over and over again, eventually it should become stronger and stable.

According to Hebb, every time we learn something new our brain will either change the structure of its neurons or increase the number of synapses between our neurons allowing them to send and receive information faster. As life proceeds, some connections and pathways get stronger, new ones are formed and some get weaker. In other words, the brain has the ability to rewire itself and unlearn by pruning unused connections, and build new ones as it learns and adapts to environmental stimulation.

We can liken this to walking through a wheat field. The first time is hard work. The second time it is a little easier, and by the time it has been done numerous times, there is a clear pathway that can be crossed effortlessly. Hebb's theory has often been summarised as 'neurons that fire together, wire together' and is still seen to be the key mechanism by which we learn and form habits and memories. In effect, Hebb was proposing that the efficiency of a synaptic connection would increase and become more effective the more often it was used. Somewhat surprisingly, even with all the evidence, it still took more than thirty years for the idea of the plasticity of the brain to become widely accepted in education and psychology.

Hebb's notion that learning occurs because of changes to the synapses was one of the biggest breakthroughs in our understanding of the brain. If nothing else, Hebb was a conscientious and committed researcher. He kept rats as pets in his house who were allowed to roam at will. On examination, he found that these rats showed greater learning ability than those kept in a laboratory. Other scientists subsequently repeated Hebb's experiment in more controlled conditions. A particularly important follow-up study came from the American psychologist, Mark Rosenzweig, and his colleagues, in 1960. They selected two groups of rats and reared them under impoverished and enriched conditions respectively. They found that rats reared in a barren, impoverished and unstimulating surroundings had fewer synapses, smaller brains and lower levels of the neurotransmitter acetylcholine. However, rats raised in an enriched environment cage filled with toys, and the company of other rats had grown longer and more dendrites, remodelled synapses and more neuronal connections. In other words, the way the brain developed was hugely dependent on experience and, significantly, the hard-working parts of the brain do indeed grow more connections. Neuroscientists refer to this as an 'enriched environment' (Rosenzweig 1966: 322).

The Power of Plasticity 59

Enriched environment

Appearance of nerve cells in the mouse brain

Standard environment

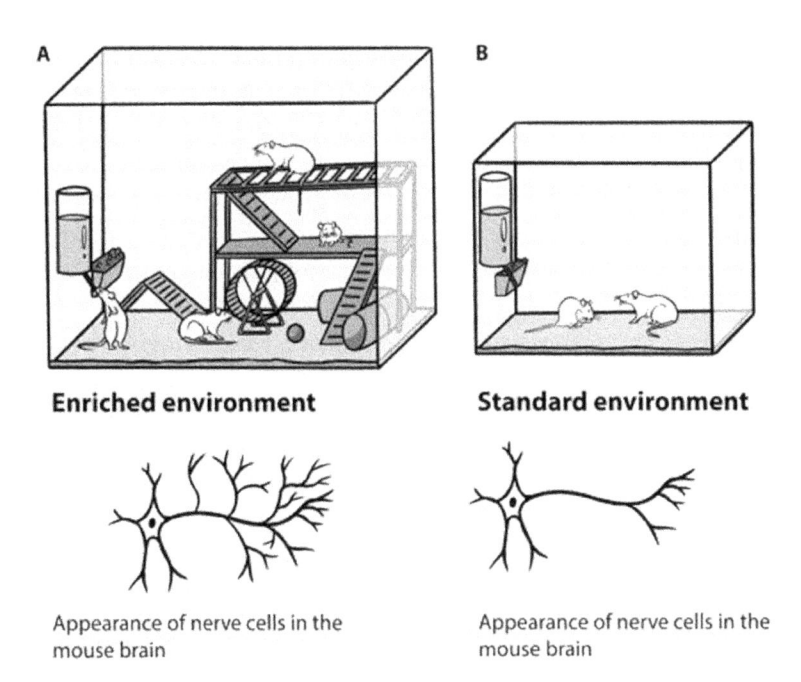

Appearance of nerve cells in the mouse brain

Figure 10: Standard vs Enriched Environment – Image by Vive et al., Attribution-Non Commercial-No Derivatives 4.0 International (Available here)

In response to this discovery, neuroscientists raised an obvious question: Is there any evidence to demonstrate that similar changes occur in the human brain? To answer this question, find out what London taxi drivers, a religious order and Leonardo DiCaprio have in common in the following case studies.

Case Study A- London Taxi Drivers

To become a London taxi driver, an applicant must undergo four years intensive training to pass the 'Knowledge of London' exam, an exceedingly difficult task. The exam requires aspiring cabbies to learn the city's geography by memorising London's extensive roadways, in all their combinations and permutations. The exam covers 320 different routes through the city, 25,000 individual streets and 20,000 landmarks and points of interest, such as hotels, theatres, restaurants, embassies, police stations, sports facilities, etc. The knowledge exam is one of the most difficult feats of learning and students typically spend 3 to 4 hours a day studying theoretical journeys.

The mental challenge of the 'Knowledge' sparked the interest of the Irish neuroscientist Professor Eileen McGuire and her colleagues at the University of London, who subsequently scanned the brains of cab drivers. The scientists were interested in the hippocampus part of the brain. They discovered that this area had grown physically larger than those in a control group (non-taxi drivers). The researchers also found that, the longer a cabbie had been doing the job, the bigger the change in the hippocampus brain region. This indicated that the growth of neurons in the hippocampus area resulted from practice. Fascinatingly, when they cease to use the 'Knowledge' regularly, their hippocampi revert back to the same size as before. The cab-drivers study demonstrates that adult brains are not fixed in place, but instead can be rewired and reconfigured. Clearly, the adage 'Use it or lose it' is as valid for the brain as for body muscles (McGuire 2000).

Case Study B – Nuns in the USA

Where can you find a stable group that can be easily tracked down each year for regular tests? Where can you locate a group with fewer 'confounding factors', or differences, that might arise in the wider population (like socioeconomic status, family structures, education levels) – all of which could interfere with research results? Where can we identify a group of people who share a similar lifestyle, including nutrition, work and living standards? Researcher's identified religious orders, because of their regulated lives and copious records as ideal subjects for brain research.

Since 1994, neuroscientists have been working with around 650 nuns from 7 different Notre Dame Convents across the US. To date, the Nuns' Study Project has collected 350 brains. Each one is carefully preserved and examined for evidence for age-related brain diseases. Over time more in-depth data is gathered on each participant during their lifetime. Each year, everyone undergoes a battery of genetic, medical, physical and psychological tests, with any work changes or career moves also monitored.

When the team began their research, they expected to find a clear-cut link between cognitive decline and causes of dementia: Alzheimer's, stroke and Parkinson's. Instead, based on autopsies on nuns' brains, they found that some suffered with the ravages of Alzheimer's disease, even though this had never been obvious to their colleagues. Some nuns were dying with a full-blown Alzheimer's pathology without having cognitive loss.

So what was going on? The team went back to their substantial data for clues. Two things became clear. Firstly, when they examined the nuns' letters of application to join the 'religious life', they observed that the applicants who expressed optimism, curiosity and adventure tended to live longer. Secondly, they found that lifestyle factors determined whether there was loss of cognition. They called this 'Cognitive Reserve Theory' – the idea that keeping cognitively fit throughout life may play a role in building resilience to help delay or prevent early degenerative brain disease.

The term 'cognitive reserve' refers to methods that boosts brain function, including education and various forms of mental training (Snowdon 2022). We will return to the concept in later chapters, but for now it is useful to think of it as a toolbox. If it is a good toolbox, it will contain all the tools you need to get a job done. If you need to disengage a screw, you might fish out a screwdriver; if you don't have access to a correct screwdriver, you may use a pair of pliers or wrench. Specifically, it appears that cognitive exercise and continuous learning throughout life, plus regular physical activity, strong social networks and having responsibilities for others play a key factor in brain health (Bennett 2012).

Case Study C – Leonardo DiCaprio

Elaine Fox in her book *Rainy Brain, Sunny Brain: The New Science of Optimism and Pessimism* relates the following story about the actor, Leonardo DiCaprio, to explain the power of brain plasticity. The American aviator, engineer and industrialist, Howard Hughes, suffered all his life from a severe disorder known as obsessive-compulsive disorder (OCD). Thirty years after the death of Hughes, the actor Leonardo DiCaprio played him in the movie *The Aviator*.

To fully engage with the role, DiCaprio spent several days with psychiatrist, Jeffrey Schwartz, to learn more about OCD. DiCaprio also spent time with some of Schwartz's patients so that he could see up close what it is like living with the illness. So immersed in the role did he become that DiCaprio developed many of the thoughts, feelings and symptoms of OCD. Interestingly, a transient case of OCD had been induced in his own brain. It took almost three years of intense therapy and practice to rid himself of OCD after filming had stopped (Fox 2012). The study of the brain of London taxi drivers, a religious society of nuns and Leonardo DiCaprio demonstrates that it is possible to protect our brains, and to help hold on to our identity for as long as possible. We can't stop the process of ageing, but by practising all the skills in our cognitive toolbox, we may be able to slow it down.

> **YouTube Video**
>
> In this video on Neurons and Learning, Professor Paul Howard-Jones talks us through what happens within the brain when we learn. Through a process of plasticity, more specifically synaptic-plasticity, the brain creates new and stronger neural pathways when we learn and connect information. All forms of learning rely on our brains to change their synapses.
>
> Available from: <https://www.youtube.com/watch?v=FTe_NPq93FY> (3 minutes)

Three Types of Plasticity

Neuroplasticity explains how the brain has the ability to continuously make new connections (neural pathways) and how old ones die out. In a nutshell, the brain has the capacity to rewire itself, pruning unused connections, and building new ones as it learns, adapts and generates new lifestyle habits. Neurons, and connections of neurons, respond to what we do and think, resulting in real changes in the way that brain circuits operate. We now know that if we change our thinking, we can also reshape our brain. However, this process is not that simple, as there is a distinction between a chemical, structural and functional change in the brain.

Chemical Change

The brain does not just change: it is constantly changing. All we have to do is chat to our friends or work colleagues over coffee, watch a documentary, visit someplace new or step outside our comfort zone into our stretch zone every once in a while. Every time we have a new experience, the brain triggers a rapid chemical change, whereupon neurons fire and connect. However these changes appear quickly and lead to changes in short-term memory. But these changes fade almost as quickly as they occur. Our short-term memory can only retain information for about 7 seconds. The next day we are pretty much back to square one. This is what happens when we do something once – there won't be any lasting change in our brains. To move to the next level of brain change, we must move out of our 'comfort zone' and deliberately and consciously set out to learn something new.

Structural Change

Structural change is the most extensively studied and best understood type of plasticity. Repeat an action several times over a period of time and that is when we see structural changes in the brain. When we mentally move from our comfort zone to our stretch zone that is when neural dendrites grow and new connections are made, which leads to long-term improvement in learning and memory.

However, this structural change takes time, it does not happen overnight and it needs repetition over and over again. Structural change in the brain comes about when we struggle to change a habit. This process is known as tuning and pruning.

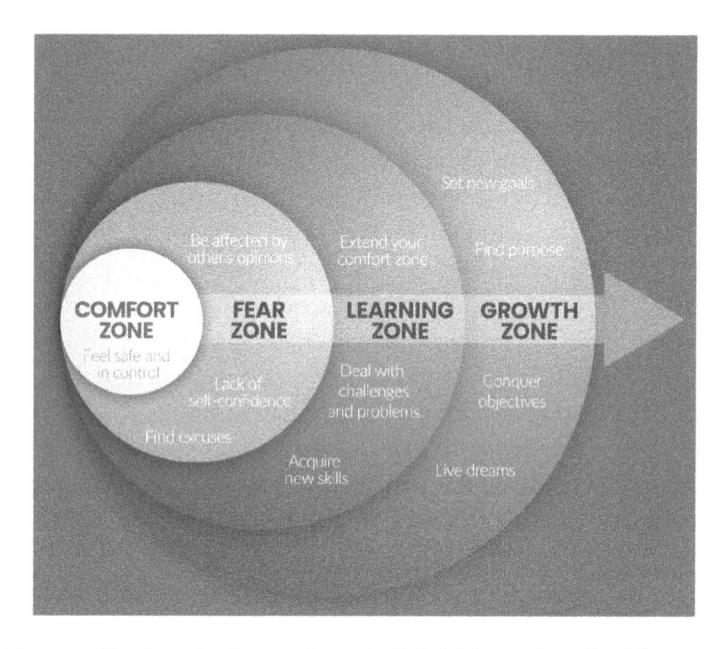

Figure 11: The Learning Zone – Image by Whale Design/<stock.adobe.com>

Tuning means strengthening the connections between neurons, particularly connections that are used frequently. If we think again of neurons as trees, tuning means that the branch-like dendrites become bushier. It also means that the trunk-like anon develops a thicker coating of myelin, a fatty bark that's like the insulation around electrical wires, which makes signals travel faster. Well-tuned connections are more efficient at carrying and processing information than poorly tuned ones and are therefore more likely to be reused in the future.

Through repetition, the brain is more likely to recreate certain neural patterns that include those well-tuned connections. Meanwhile, less used connections weaken and die off. This is the process of pruning, the neural equivalent of 'If you don't use it, you lose it', as we noted earlier. Pruning is critical in a developing brain, especially synaptic pruning. Just like gardeners with trees and bushes, the brain has its own system of cutting off unused connections, and, at the same time strengthening those that are in regular use (Magrini 2019).

Practising a new activity helps make us better at it – repeated activities strengthen interactions between the relevant neurons inside the brain. After years of practice, the activity become so engrained in the neural networks that it become 'automatic', thus leaving more space in the brain for new learning (Brennan 2019). As we initially fumble around, trying to retain new information or acquire a new skill to have it 'hard-wired' into the brain the senses are sending repeated signals coursing through a specific pattern of neurons. Remember, for example, when we first started to drive. After one lesson, little will probably occur in the brain to keep that knowledge or skill intact. But after continued firing and practice a critical threshold is eventually reached and the neurons physically change. All our habits are acquired and developed in this way (see Chapter 7).

The 10,000 hour rule, introduced by Anders Ericsson (1993) but made popular by Malcolm Gladwell in his book *Outliers (2008)*, suggests that becoming an expert takes 10,000 hours of practice. While his concept has been challenged by scientists (Mcnamara et al. 2016), and rebutted by Ericsson (2013) who described the concept as 'a provocative generalisation'. However, there is some truth to the notion that it takes many years of practice (repetition) to change the brain's structure. Importantly, the age at which someone gets involved in an activity seems to play a role in their ability to achieve mastery (Gobet and Campitelli 2011). As with language learning, there may be a window during childhood when specific, complex skills are most easily acquired. Genetics play a role as well, a study of 10,000 twins conducted by Mosing et al. (2014) in Sweden found that genes influenced 38 per cent of the musical abilities they measured.

Functional Change

The third and last type of change in the brain is functional change. This is the most impressive form of plasticity when we see whole parts of a brain that were allocated to doing one thing and then starting to change to do other things. This happens frequently – for example, when someone has a stroke and they lose the ability to use one hand. In previous times, if a person had an injury in their right hand, they would be helped to develop the function of their left hand. This is not the modern way of reacting to a stroke. Because of plasticity, a person recovering from stroke would be encouraged to force the brain to find new pathways towards partial or full recovery of the right hand.

Neuroplasticity can thus come to our aid in response to injury and illness. Perhaps one of the most incredible examples of this is the case of US Representative, Gabrielle ('Gabby') Giffords, who, on 8 January 2011 was shot during a meeting held in a supermarket parking lot in Tucson, Arizona. Gabby was shot in the back of the head with a 9 mm bullet at point-blank range. The bullet travelled through her brain, exiting through her forehead. It was the type of injury that few survive. Fortunately, for Gabby, she did, because as the bullet passed through her left hemisphere it only damaged her speech and language production part of her brain. This enabled her medical team to make the most of her brain's inherent neuroplasticity by encouraging her to sing rather than attempt to talk as part of her rehabilitation (Doidge 2007). As the brain's right hemisphere is recruited more for music than for language, they engaged different neural networks and over time enabled her to regain many of her previous abilities (Loveday 2016).

Gabby's story is a powerful illustration of how the brain compensates for damage by forming new communications between interacting neurons. It is a reminder too of its incredible capacity for growth, protection and resilience. Such functional brain plasticity may take many years with thousands of hours of hard work and patience. There are many such incredible stories of plasticity in the literature.

Conclusion

In this chapter, we have outlined the discovery of the neuron and how developments in brain science have led to deeper understandings of neuroplasticity – that is, that the human brain can be modified as a result of experience. Evidence from novel studies are presented in this chapter in the form of case studies, such as the nuns and London taxi driver studies, to demonstrate how the brain is modified by experience. Three types of plasticity are presented, chemical, structural and functional. Of particular relevance to education is the structural dimension, which explains the physical changes that take place in the brain when we learn.

Neuroplasticity is a revolutionary idea which has an impact on how we think about learning, and has implications for how we teach and how we define intelligence. It enables the brain to encode new information and skills. When we learn something new, whether it's a fact, a motor skill, or a new language, the brain forms new connections or strengthens existing ones. The complex networks of neurons and pathways of nerve fibres inside our heads constantly respond, adapt and rearrange themselves. This malleability presents us with a fantastic opportunity to take control of our own brain.

References

Bennett, D. A., Schneider, J. A., Arvanitakis, Z. and Wilson, R. S. (2012). Overview and Findings from the Religious Orders Study. *Current Alzheimer's Research*, 9 (6), 628–645.
Brennan, S. (2019). *100 Days to a Younger Brain*. London: The Orion Publishing Group.
Cajal, Ramon y S. (1897). *Advice for a Young Investigator*. Madrid: MIT Press.
Craver, C. (2007). *Explaining the Brain: Mechanisms and the Mosaic Unity of Neuroscience*. Oxford: Oxford University Press.
Doidge, N. (2007). *The Brain That Changes Itself: Stories of Personal Triumph from the Frontiers of Brain Science*. London: Penguin Books.
Ericsson, A. (2013). The Danger of Delegating Education to Journalists: Why the APS Observer Needs Peer Review When Summarizing New Scientific Developments. http://blogs.ischool.berkeley.edu/i225s14/files/2014/04/2012-Ericssons-reply-to-APS-Observer-article-Oct-28-on-web.pdf.
Ericsson, K. A., Krampe, R. T., Tesch-Romer, C., Ashworth, C., Carey, G., Crutcher, R. J., Grassia, J., Hastie, R., Heizmann, S., Judd, C., Kellogg, R., Levin, R., Lewis, C. H., Oliver, W., Poison, P. G., Rehder, R., Schlesinger, K., Schneider, V. I. and Wilson, J. (1993). The Role of Deliberate Practice in the Acquisition of Expert Performance. *Psychological Review*, 100, 363–406.

Fernandez-Armesto. (2019). *Out of Our Minds: What We Think and How We Came to Think It.* UK: Oneworld Publishers.
Fox, E. (2013). *Rainy Brain, Sunny Brain: The New Science of Optimism and Pessimism.* London: William Heinemann Publishers.
Gladwell, M. (2008). Outliers: The Story of Success. New York: Little, Brown & Co.
Gobet, F. and Campitelli, G. (2011). Deliberate Practice: Necessary But Not Sufficient. *Current Directions in Psychological Science*, XX (X), 1–6.
Hebb, D. O. (1949). *The Organization of Behaviour: A Neuropsychological Theory.* New York, NY: John Wiley.
Kuhn, T. (1996). *The Structure of Scientific Revolutions.* Chicago: Chicago University Press.
Loveday, C. (2016). *The Secret World of the Brain: What It Does, How It Works and How It Affects Behaviour.* London: Andre Deusch Limited.
Macnamara, B. N., Moreau, D. and Hambrick, D. Z. (2016). The Relationship between Deliberate Practice and Performance in Sports: A Meta-analysis. *Perspectives in Psychological Sciences*, 11 (3), 333–350.
Magrini, M. (2019). *The Brain: A User's Manual: A Simple Guide to the World's Most Complex Machine.* London: Short Books.
Maguire, E. A., Gadian, David G., Johnsrude, Ingrid S., and Frith, Christopher D. (2000). Navigation-Related Structural Change in the Hippocampi of Taxi-Drivers. *Proceedings of the National Academy of Science of the United States of America*, 97 (8), 4398–4403.
Mosing, M., Madison, G., Pedersen, N., Kuja-Halkola, R. and Ullén, F. (2014). Practice Does Not Make Perfect: No Causal Effect of Music Practice on Music Ability. *Psychological Science*, 25 (9), 1795–1803.
O'Shea, M. (2005). *The Brain: A Very Short Introduction.* Oxford: Oxford University Press.
Rosenzweig, M. R. (1966). Environmental Complexity, Cerebral Change, and Behaviour. *American Psychologist*, 21 (4), 321–332. <https://doi.org/10.1037/h0023555>.
Rosenzweig, M. R. (1979). *Development and Evolution of Brain Size*, 263–293. In Hahn, M., Jensen, C. and Dudek, B. (Eds). Development and Evolution of Brain Size: Behavioural Implications. Boston, MA: Academic Press.
Snowdon, D. (2002). *Aging with Grace: The Nun Story: Living Longer, Healthier, and More Meaningful Lives.* London: Bantam Books.

SECTION 2

CHAPTER 5

What Is Educational Neuroscience?

> 'Education is about enhancing learning and neuroscience is about understanding the mental processes involved in learning. The common ground suggests a future in which educational practice can be transformed by science'.
>
> (Blakemore et al. 2011: v)

Educational neuroscience is an interdisciplinary research field that bridges the gap between neuroscience and education. It is an emerging field whose goal is to translate new insights from the study of neural mechanisms underpinning learning, into practical applications in the classroom to improve educational outcomes (Thomas and Ansari 2020). Drawing on research in behavioural and cognitive psychology, as well as neurobiology, it translates research on the neural mechanism of learning to educational practice and describes how learning changes the brain (Thomas et al. 2019).

In this chapter we will introduce the key thinkers and concepts which laid the foundation for the emergence of contemporary educational neuroscience, and consider their role in understanding the link between learning and brain health.

Biological Foundations of Learning

The relevance of neurobiology to education has long been recognised. Our initial understanding of non-conscious learning came from behavioural psychologists who focused research on the gradual learning and unlearning of habits, or the conditioning of the learner. Beginning with educational psychology in the early 1900s, Thorndike and Dewey became leaders in shaping new approaches to learning, drawing on the influence of physical and biological sciences, which ultimately led to the establishment of educational neuroscience. Despite being colleagues in Teachers College in the early twentieth century, Dewey and Thorndike spearheaded different approaches.

Edward Thorndike laid the foundation for understanding behaviour and learning. He believes learning is the result of trial and error. He discovered from his experiments that reflection, memory, background experiences and intelligence have nothing to do with the learning process. He formulated this notion into the theory of the Law of Effect. These laws include three major elements of learning: the stimulus element which involve environmental event, the response element that include behavioural acts and the formation of a connection or bond. The principles of learning can be used to teach children desirable behaviours, like non-disruptive classroom conduct.

The Law of Effect states that behaviours that are responsible for producing positive outcomes will be repeated, whereas, those which are not, will not be repeated. Rewards increased the strength of connections between stimuli and responses, therefore while learning happens through trial and error, learning is more likely to occur where the consequences are satisfying. It is Thorndike who identified these connections between stimulus and response as being represented in a neural bond.

Thorndike's work was expanded upon by B. F. Skinner (1954) who coined the term 'operant conditioning', his main thesis being that actions that are followed by reinforcement will be strengthened and more likely to occur again in the future. While these early theories of the law of effect, and operant conditioning, have been replaced by more current theories that consider higher order and complex forms of learning (Schunk 2016: 88), understanding the importance of the reward process in learning is still useful. When we understand the basic principles of operant conditioning, we can design and develop a new habit. When it comes to learning it becomes a matter of breaking the habit down into its individual parts, as we discuss in Chapter 7.

John Dewey believed that humans develop behaviours in response to their experiences and interactions with the environment. He emphasised the importance of shaping the educational environment to promote active inquiry as a means of learning by doing. He experimented with educational curricula, arguing that traditional subjects are important and learning should not be entirely led by the child's interests. He advocated for methods which require learners to engage in defining problems, testing ideas and finding solutions, claiming that we do not learn from experience. Rather we learn from reflecting on experience (Dewey 1910). Dewey argued that collaborative or shared learning is important for development because it induces an emotional response which results in deeper learning. His ideas that linked learning and emotion are central to our current understanding of how the brain consolidates learning.

Socio-cultural Foundations of Learning

Jean Piaget, a Swiss psychologist, with an interest in sociology and biology, challenged the traditional view of learning and cognition, and inspired several generations of constructivist educationalists who followed. Piaget (1964) claimed that a stimulus-response model of behaviour could not explain cognitive learning. Studying young children's learning, he suggested that children do not internalise knowledge and skills that are presented by others, or through rote learning; instead, children construct their own knowledge based on their own cognitive and physical development. The learners' environment and how they internalise and respond to external stimulus therefore plays a central role in cognitive and brain maturation.

Piaget's theory of 'stages of cognitive development' claims that children go through stages of learning in response to the way they adapt to their environment. Children interact with their environment and as they do, they learn to create knowledge – initially processing concrete information (explored through the senses), then advancing to more abstract concepts like object permanence (e.g. other people are separate to me). In this sense, learning can be seen as a holistic process influenced by the body and mind through mostly spontaneous interaction with the environment. Piaget conducted experiments to test this theory that learning is possible when relationships and structures develop naturally, rather than through external reinforcement or instruction (Zhang 2022). His work contributed to understanding the gradual growth and maturity of the brain of the learner, which is indicative of the gradual maturity of the brain as we move from childhood to adulthood. Educational neuroscientists build on Piaget's work by associating biology and environment with brain development.

Vygotsky criticised Piaget's theory for not addressing non-spontaneous learning and the influence of the social context on learning. Another key player in the development of cognitive psychology, Vygotsky's socio-cultural theory emphasised the importance of social interaction and instruction for enhancing learning. Vygotsky's theories stress the fundamental role of social interaction in the development of cognition and the process of 'making-meaning'. Each individual is shaped by, and shapes the people around them, with adults (as more knowledgeable others) playing a particularly important role in engaging and challenging children in meaningful ways. His theory of the Zone of Proximal Development (ZPD) proposes that it is through the support of knowledgeable others that learners expand their learning and cognition. The ZPD refers to the space or zone between what a child is capable of doing on their own and what they can achieve with assistance from an adult or more capable peer. Mental functions such as attention, memory and perception develop and become 'higher mental functions' through interaction with the sociocultural environment and the more knowledgeable other.

A challenge faced by all educators is the clear differentiation between lower-order and higher-order practices. In this context, one of the more enduring and useful models for enhancing thinking was developed by Benjamin Bloom in the 1950s. Bloom's classification or taxonomy is considered one of the most useful tools for moving students to higher levels of thinking. In a number of landmark papers, Bloom and colleagues identified three learning domains: the cognitive, the affective and the psychomotor. The cognitive domain is the one that concerns us here as it involves thinking of all sorts at six levels, from the least to the most complex. They are as follows: knowledge, comprehension, application, analysis, synthesis and evaluation. Although they are six separate levels, the hierarchy of complexity is not rigid, and the individual may move easily among the levels during extended processing.

According to Bloom, the acquisition of facts (knowledge) marks only the beginning of understanding. The facts must be understood (comprehension) before they can be applied to new situations (application). Knowledge must be organised and patterns recognised (analysis) before it can be used to create new ideas (synthesis). Finally, to discriminate among competing models or evidence, the learner needs to be able to assess (evaluation) the relative merits and validity of information or ideas. By systematically following teaching practices with a higher-order level focus, we can foster novelty-seeking creative brains to grow towards higher levels of evaluation and creativity.

Cognitive Psychology and Neuroscience

Educational neuroscience, as we know it today, has been heavily influenced by cognitive neuroscience, that is the interdisciplinary field bringing together cognitive psychology (the study of instruction) and neuroscience (the study of the brain). While cognitive psychology is not typically interested in the study of the structures or functions of the brain, cognitive neuroscience uses cognitive models and brain recording techniques to examine how neural circuitry changes and develops due to mental functions of the brain.

Cognitive psychology, known as the science of learning, has contributed to the design of educational approaches. Jerome Bruner (1966), a cognitive psychologist who was influenced by Piaget, Vygotsky and Skinner rejected the idea that belief, desire and internal mental states exist. Bruner (1947) argued that there may be an interaction between the physical processes of the body and the mental processes of the brain. He agrees with Piaget that learners move through stages of intellectual development but suggests that these stages are not necessarily age dependent. In his 1960 text Bruner, in The Process of Education, proposed that students are active learners who construct their own knowledge. Adults should facilitate learning by providing opportunities for problem solving and higher-order thinking. By revisiting concepts, by making connections with existing knowledge and by practice, we can learn anything at any age. As he states, 'We begin with the hypothesis that any subject can be taught effectively in some intellectually honest form to any child at any stage of development' (1960: 33). Subjects should be taught at levels of gradually increasing difficulty so that our brains can process and store the information in different ways.

In his 'discovery learning' model, Bruner identifies three ways that information is processed. Firstly, in 'enactive mode', learning and thinking based on action. The second stage, 'iconic mode', is when information is stored as sensory images, like visual diagrams or images. Finally, the third stage is 'symbolic mode', where knowledge is stored as language or symbols. His theory requires the learner to have autonomy, and for a more knowledgeable person to scaffold the learning experience. He suggests that the teacher or adult should provide only partial solutions, encouraging trial and error, motivating the learner to keep going, and ensuring that the learner is clear on both problem and task. Lessons should be taught at levels of gradually increasing difficulty so that our brains can process and store the information in different ways. This emphasis on how knowledge is stored, is one of the central themes in educational neuroscience.

The Emergence of Educational Neuroscience

The birth of educational neuroscience can be traced back to developments in neuroscience and cognitive neuroscience. From the inception of CT brain imaging in the mid-1950s, MRI technology in the late 1970s, and fMRI resonance imaging from 1990s, neural imaging technology has allowed us to map the brain, and help us to explain how chemical and electrical signals produced by neurons interact with cognitive and mental processes like memory and language. These discoveries led to an increase in interdisciplinary research and encouraged joint contributions between neuroscientists and cognitive psychologists. In the twenty years that followed, questions examined by cognitive neuroscientists, particularly in relation to memory, language, attention and the brain began to permeate the educational field. In particular, there was a growing interest in neurodiversity of cognitive outputs, or learning disorders (Dundar et al. 2016). The CRiSTLE report (Case Study), provides further reading on the application of neuroscience to education.

> Case Study
>
> *The Learning Sciences in Teacher Education* project was developed by Bath Spa University's Centre for Research in Scientific and Technological Learning (CRiSTLE) was funded by The Wellcome Trust. The project created resource materials to engage trainee teachers in primary initial teacher education (ITE), and aimed to give trainee teachers confidence in being able to read, understand, critique and apply relevant scientific research to their teaching. McMahon et al. (2020),
>
> Available from: <https://www.bathspa.ac.uk/projects/learning-sciences-in-teacher-education/>

Unlike cognitive neuroscience, educational neuroscience is focused on understanding how cerebral processes affect learning (and subsequently teaching). Educational neuroscience therefore draws on insights from psychology (predominantly cognitive psychology), neuroscience and pedagogy to examine the interaction between learning and brain science. It wasn't until the 1990s, coined 'The Decade of the Brain' by the US Congress, that a movement entitled 'brain-based learning' attempted to explicitly link neuroscience and education, thus gaining greater widespread visibility for neuroscience (Jones and Mendell 1999).

In 1997, Bruner's seminal article explained that the links between neuroscience and education rely on three basic premises. Firstly, that there is a dramatic increase in the number of synapses that connect neurons in the brain as a child develops. Secondly, sensory and motor systems are influenced by one's experiences. And thirdly, complex or enriched environments cause new synapses to form, meaning our brain can continue to develop (Bruner 1997: 4). However, at that stage, he argued that there was not enough evidence to link neural development with educational practice.

Learning how to learn is the modern focus of education. Using data from neuroscience, we can take a broader perspective on educational outcomes, taking into consideration the impact of sleep, diet, stress and exercise. What these debates highlight is that the 'translation' of knowledge from neuroscience to education, and one's engagement with research in the area, must be approached with a degree of criticality and awareness of the field. As such, this current chapter will discuss the history of educational neuroscience and the key thinkers and concepts relevant to the interdisciplinary study of how learning happens and what happens in the brain when we are learning.

If scientists and lifelong educators can better understand how the brain functions and how learning happens, then they are better equipped to facilitate students to harness the knowledge of how to learn, and ensure lifelong learning in their students. Recent research tells us that teaching students about brain plasticity can actually have a positive effect on their motivation and their attainment, and that learners who understand brains as 'plastic' are more resilient (Paunesku et al. 2015; Dubinsky et al. 2013). However, differences in terminology, language and approaches have meant that knowledge from neuroscience has not always been accurately translated in education (Howard-Jones 2014: 817), leading to an industry founded on neuromyths such as preferred learning style or right-brain, left-brain learners. Such myths over-simplify learning.

In a critique of educational neuroscience, Bowers (2016) argues that neuroscience adds nothing more than psychology to our understanding of learning, whereas Howard-Jones et al. (2016: 620) feels this is a step too far, describing it as 'potentially misleading' arguing that psychological and neural levels of explanation complement rather than compete with each other. The brain is a biological organ that requires optimal condition to function, particularly to learn. Therefore we must acknowledge the link between brain health and learning as there is a link between neuroscience and education as well as indirect links between neuroscience, psychology and education, there are also more direct links between neuroscience and the capacity to learn. What makes us unique is a complex combination of factors – our lifestyle and our habits (fitness, diet, sleep); the social, economic and cultural factors that constrain us (air quality, educational experiences) as well as biological and psychological factors that make us unique (motivation, genetic make-up, stress). These all impact on our brain function and we bring these with us to the learning environment (Thomas et al. 2019).

Since the 1990s, the growth in the field of educational neuroscience has led to the development of various different research groups and centres who focus explicitly on furthering research into brain functions or dysfunctions; for example, in 2009, the European Association for Research on Learning and Instruction (EARLI) established a special interest group called 'Neuroscience and Education'. Cambridge University and University College London founded the 'Centre for Neuroscience in Education' (2005) and 'the Centre for Educational Neuroscience' (2008) respectively. In 2022, the Organization of Economic Co-operation and Development's (OECD's) New Approaches to Economic Challenges Unit (NAEC) created the Neuroscience-inspired Policy Initiative which acknowledges the importance of brain health for productivity, social reality and quality of life. According to the OECD, 'We live in a brain economy that places a premium on creativity and intellectual skills' (2011: 1431), and yet we sometimes fail to make the connection between brain health and productivity. At local level, organisations and programmes are integrating findings from Neuroscience into Education Coaching (Positive Success) and teachers' manuals are being created to try to encourage educators to consider the links between brain functions and learning.

Expert Insight:

Dr William Kitchen, Stranmillis University College

In conversation with the authors, Dr Kitchen, author of *Philosophical Reflections on Neuroscience and Education*, highlighted the importance of how we approach the application of neuroscience research in education.

Neuroscience is not a silver bullet for teaching and learning

'Recent developments in neuroscience around the way the brain functions, are actually starting to shed light on so called educational myths. We're are all desperate to try and find better ways to educate the children in front of us because we genuinely care about doing a better job for them. When someone comes with a silver bullet idea, and it is in some way backed up by ideas about the brain, then my goodness it appears that it's going to solve all problems. The temptation for us as teachers is to think that if we know more about the brain, then we know more about the child'.

'There are certain mythologies that have become deeply embedded in educational psyche. They are very difficult to shift because teachers have embedded them into their practice. Advancements, particularly because of the technological advancements that we now have like fMRI, are actually starting to show us that the prevailing orthodoxy in these wisdoms that we thought were true. The notion of 'learning styles' is a perfect example. Or the suggestion that there was a causal connection between little areas underneath the ribcage and the ability to do Maths. When children were doing mathematics, they were told they had to stand up and poke and underneath their ribcage because they thought it might make them better mathematicians'.

'To remove those mythologies, we actually have to use very sophisticated examples of neuroscience to demonstrate that they are unfounded. Neuroscience has started to lend itself to types of movements, particularly things like growth mindset, and computational thinking. People are looking for scientific credibility, to give more strength to an idea that may well otherwise be resisted by the educational establishment, so people are tending to dip into neuroscience more regularly. Those attempts at linkages give an insight into the way the whole area of mind-brain-education or neuro-education has developed. Those of us who are genuinely interested in the application of neuroscience to education, have a responsibility to try and find a pathway for taking the pieces of neuroscientific evidence that are of profound significance and value to our practice; but also making sure that we don't overvalue it to the point that we feel we've cracked it. Social issues, economic problems, interpersonal problems will still remain'.

We need both education and neuroscience, not either or

Neuroscience to me has to be the start of a conversation rather than the handoff.

> Things that are inherent in learning are not just about what goes on inside the brain. Learning is very often about the cajoling that we've had to do to even get the child to open the book, the arm around the shoulder when we think the child is struggling ... those things don't take place in the brain, they take place in the social interaction that takes place between teachers and pupils, and indeed, between pupils and pupils. Knowing more about the brain can give us insights into things that we would want to know more about. Neuroscience and Education have been used effectively to support the field of Adverse Childhood Experiences (ACE). Say that we come across a child in the classroom who genuinely does not appear to have the capacity to regulate their emotion. They may look as though they just get angry, because anger is the only emotion that they can experience. When we look at what some neuroscientists say about this we may discover that the amygdala does not function the way that a normal amygdala would function. It doesn't function in the way that it allows the child to regulate their emotions. The child therefore doesn't have the capacity to regulate emotions in the way that we might expect a child to do in other circumstances. Neuroscience is shedding an entirely new light on what way we view behaviour, and providing evidence to support the idea that behaviour is communication for something deeper. Similarly, neuroscience shows that brain development can be hindered by career and life choices. We can start to open that conversation a little bit more with children and young people about the types of impact that negative choices in their lives can have for their capacity to do things later on in life'.
>
> **Recent Research**
>
> Kitchen, W. H. (2021). Neuroscience and the Northern Ireland Curriculum: 2020, and the Warning Signs Remain. *Journal of Curriculum Studies*, 53(4), 516–530.
> Kitchen, William H. (2017). *Philosophical Reflections of Neuroscience and Education*. New York, NY: Bloomsbury Academic.

Conclusion

Learning is commonly understood as the process through which human beings acquire or modify their ability, skills and knowledge or behaviour as a result of studying, observing or teaching (Şentürk and Baş 2020). This chapter has introduced critical insights from the field of neuroscience, ranging from the biological foundations of learning, socio-cultural learning, cognitive psychology and the more recent emergence of educational neuroscience. The case studies and vignette in this chapter provide interesting insights into recent research in the field. The key message for students is that learning *how* to learn, rather than *what* to learn is crucial.

There is no doubt that educational neuroscience has potential to inform educational policy and practice by providing empirical evidence for the effectiveness of specific teaching strategies, curricula and interventions. As Dr Kitchen highlights, neuroscience research can be effectively used to shed light on education practice. However he cautions that we must engage critically with the research that is presented and be aware of neuromyths that persist.

References

Blackwell, L. S., Trzesniewski, K. H. and Dweck, C. S. (2007). Implicit Theories of Intelligence Predict Achievement across an Adolescent Transition: A Longitudinal Study and an Intervention. *Child Development*, 78(1), 246–263.

Blakemore, C., Frith, U., Harris, J., Mackintosh, N., Rees, G., Robbins, T., Rose, S., Sahakian, B., Singer, W., Stirling, A., Tracey, I., Schultz, W. and Chan, S. (2011). *Brain Waves Module 2: Neuroscience: Implications for Education and Lifelong Learning.* London: Royal Society.

Bloom, B. S. (1956). *Taxonomy of Educational Objectives, Handbook I: The Cognitive Domain.* New York, NY: David McKay Co.

Bowers, J. S. (2016). The Practical and Principled Problems with Educational Neuroscience. *Psychological Review*, 123(5), 600–612.

Bruner, J. S. (1960). *The Process of Education.* Boston Harvard University Press.

Dewey, J. (1938). *Experience and Education.* New York: Macmillan.

Dewey, J. (1910). *How We Think.* Boston: D. C. Heath & Co Publishers.

Dubinsky, J. M., Roehrig, G. and Varma, S. (2013). Infusing Neuroscience into Teacher Professional Development. *Educational Researcher*, 42, 317–329.

Dundar, S. and Ayvaz, U. (2016). From Cognitive to Educational Neuroscience. *International Education Studies*, 9(9), 50–57.

Eyre, H., Ayadi, R., Ellsworth, W., Argam, G., Smith, E., Dawson, W., Ibanez, A., Altimus, C., Berk, M., Manji, H., Storch, E., Leboyer, M., Kawaguchi, N.,

Freeman, M., Brannelly, P., Manes, F., Chapman, S., Cummings, J., Graham, C., Miller, B., Sarnyai, Z., Meyer, R. and Hynes, W. (2021). *Building Brain Capital*. OECD. *Neuron* 109(9):1430–1432.

Howard-Jones, P. (2014). Neuroscience and Education: Myths and Messages. *Nature Reviews Neuroscience*, 15(1), 817–824.

Jones, Edward G. and Mendell, Lorne M. (1999). Assessing the Decade of the Brain. *Science*, 284(5415):739.

Paunesku, D., Walton, G. M., Romero, C., Smith, E. N., Yeager, D. S. and Dweck, C. S. (2105). Mind-set Interventions Are a Scalable Treatment for Academic Underachievement. *Psychological Science*, 26(6), 784–793.

Piaget, J. (1964). Part 1: Cognitive Development in Children: Piaget Development and Learning. *Journal of Research in Science Teaching*, 2(3), 176–186.

Schunk, Dale H. (2016). *Learning Theories: An Educational Perspective* (6th edn). Boston: Pearson Press.

Şentürk, C. and Baş, G. (2020). An Overview of Learning and Teaching from the Past to the Present: New Learning and Teaching Paradigms in the 21st Century. In S. Orakci (2020). *Paradigm Shifts in 21st Century Teaching and Learning* (pp. 1–19). Pennsylvania IGI Global Publishers.

Skinner, B. F. (1954). The Science of Learning and the Art of Teaching. *Harvard Educational Review*, 24, 86–97.

Thomas, M. S. C. and Ansari, D. (2020). Educational Neuroscience: Why Is Neuroscience Relevant to Education?. In M. S. C. Thomas, D. Mareschal and I. Dumontheil (Eds), *Educational Neuroscience: Development across the Life Span* (pp. 3–22). Milton Park: Routledge.

Thomas, M. S. C., Ansari, D. and Knowland, V. C. P. (2019). Annual Research Review: Educational Neuroscience: Progress and Prospects. *Annual Research Review: Journal of Child Psychology and Psychiatry*, 60(4), 477–492.

Thorndike, E. L. (1926). *Educational Psychology. Volume 1: The Original Nature of Man*. PhD thesis, Teachers College, New York.

Vygotsky, L. (1978). *Mind in Society the Development of Higher Psychological Processes*. Cambridge, MA: Harvard University Press.

Zhang, J. (2022). The Influence of Piaget in the Field of Learning Science. *Higher Education Studies*, 12(3), 162–168.

CHAPTER 6

How Learning Happens

'More is not better, being better at the process is better'.

(Sullivan and Parker 2016: 246)

In simple terms, learning involves forming and strengthening neural connections and networks or synaptic connections. Building on Hebb's theory (1949), we know that we are born with a large number of synaptic connections and as we grow existing connections can be strengthened, lost or new ones formed. Therefore, learning does not occur in the brain merely when information is received; rather, information is processed from working memory (where it is received) into short term memory (where it is filtered into important or unimportant). It is only then that it is processed into long term memory, where is may be connected with existing information. When information is used, rehearsed or recalled, the neural connections increase in number and are strengthened. This means that learning is consolidated, is more easily accessible and appears to occur more automatically. Both internal and external stimuli can help the learner to strengthen connections and consolidate learning.

In this chapter, we look at the external factors or strategies which can be understood and utilised to support the brain in its efforts to learn effectively. There are many different types of learning: observational learning (Bandura 1965), habit formation (Wood et al. 2002), patterning (Caine and Caine 1994), perceptual and motor learning (Censor et al. 2016), learning of facts (Woloshyn et al. 1992), learning by making inferences (Friston 2002), all of which are complementary and interconnected in practice. Each type of learning provides unique stimuli for the brain, but the information provided is processed by the whole brain through a series of complex mechanisms. For example, the hippocampus is heavily involved in factual and rule-based learning as well as spatial navigation, but it is also centrally important for statistical learning (Schapiro and Turk-Browne 2015). The following sections will explore cognitive load, metacognition and the ways in which teaching, learning and practice strategies can be used to ensure engaged and effective opportunities for deep learning.

The Neural Mechanisms of Learning

Due to brain plasticity, our brains develop differently depending on the synaptic connections or networks that we develop through learning. Therefore learning shapes our brain as much as our brain shapes learning! Learners dynamically and actively construct their own brain networks as they navigate through social, cognitive and physical contexts. According to the Committee on How People Learn (2018: 59), 'the reciprocal interactions in learning between the dynamically changing brain and culturally situated experience form a fascinating developmental dance, which we are discovering more and more about each year'. Neural tuning happens because of experience and is part of the reason why each individual's brains develops differently. For example, the brains of advanced readers show greater specialisation for words than those of less experienced readers (Dehaene et al. 2010). Similarly, cultural experiences and interactions shape the brain's emotional and social processes (Kitayama et al. 2017).

Educationalists, depending on their ontological position, have developed different approaches to and arguments about how learning happens. In this context the interaction between fields of neuroscience and psychology, and then psychology and education has been advanced by a number of researchers (Mareschal et al. 2013; Howard-Jones et al. 2016). Crucially, this interaction requires that evidence from neuroscience be used to modify traditional theories of psychology and cognitive sciences.

As we have seen behaviouralists argue that learning happens when connections are made between stimuli and responses. Any kind of learning is assumed to be fundamentally governed by the forming of some link between a particular stimulus and a specific response. The abbreviation commonly used is S-R, stimulus response learning. The hallmark of behaviourism is conditioning theory, which explains learning in terms of environmental events. The general principle is that learning is an observable phenomenon demonstrated through learner behaviour. For example, the observer is able to display behaviour or imitate the behaviour that they have observed.

Social cognitive theorists agree that learning is largely an information processing activity but they break this down into learning that happens through 'doing' or learning that happens through 'observing' (Bandura 1986). This is sometimes described as implicit and explicit learning. Explicit learning describes what we read, write and talk about, while implicit learning is what we learn about through experience. The more complex the knowledge or skill being learned, the more likely it is that both observation and action must come together to ensure and consolidate learning. For that reason, modelling is key to social cognitive theory. According to Bandura (1977), observational learning occurs through modelling when the learner (a) pays attention to a particular event or behaviour, (b) they store the information in memory, (c) they rehearse the knowledge by coding it or making connections and (d) they produce knowledge which involves translating information into action. Unlike behaviourists, social cognitive theorists acknowledge learner individuality and interpretation.

Information processing theorists such as Atkinson and Shiffrin (1971) claim that forming associations between stimuli and responses does help the learner to store knowledge in memory, but they are more interested in the internal processes undertaken by the brain during the learning process. They believe that the human brain functions like a computer, in that cognitive processing is efficient and there is little waste. Learning engages the whole physiology. We know from recent neuroimaging that the amygdala and prefrontal cortex cooperate with the medial temporal lobe in an integrated way in order to enable consolidation of learning. The prefrontal cortex engages in encoding and forming memory, while the hippocampus successfully stores the information in the long-term memory (Tyng et al. 2017).

Neuroscience theories can provide educators with a theoretical lens through which to understand, examine and explore the relationships between student learning, teaching and their own pedagogical practice

(Kullberg et al. 2017). However, as Howard-Jones (2014) says, classroom-ready knowledge from neuroscience is never likely to be easily transferable. We do not purport to provide instruction for learners and teachers on how to do 'brain-based' teaching. Rather we present a synthesis of research which allows us to understand why some brain compatible techniques may work better than others.

> ### TED Talk
>
> Daniel Kahneman (2010) asks why do we put so much weight on memory relative to the weight we put on experience? In his TED talk on *The Riddle of Experience vs Memory* in Longbeach California, Kahneman says that one of the traps about thinking about happiness is confusion between experience and memory – being happy in your life or being happy with your life. There is an experiencing self who lives in the present; and there is a remembering self, is the one that maintains the story of our life.
>
> Available from: <https://www.youtube.com/watch?v=XgRlrBl-7Yg>

The Importance of Attention, Engagement and Cognitive Load

Engagement is not a requirement of all learning, as we have all experienced unconscious learning or 'unconscious acquisition' when we have been out for a walk or when doing the shopping. Probably more than 90 per cent of what we learn is as a result of unconscious learning (Gazzaniga 2001), but engagement is very important for priming of explicit learning. Priming requires us to ignore external and internal distractions and to pay attention using sight, sound and physical posture. When we pay attention, our prefrontal cortex inhibits distractions, our reticular system is ready for new information, our thalamus categorises new information and our pulvinar nucleus establishes focus and our posterior parietal lobe disengages. Commanding a learner's attention can be a challenge for both the teacher and the learner, as we have limited capacity for attention.

Within the brain, the first sensory intake filter is the Reticular Activating System (RAS), which consists of a network of cells in the lower brain stem through which sensory input (information or stimuli) must pass. It is essentially the gatekeeper of the brain and of all the inputs that a learner receives each second (auditory, visual, etc.), only a fraction gets through the gates. Sights, sounds, sensations and feelings bombard us in social situations and often prevent us from focusing completely on a task. Deciding on what to foreground and what to disregard can be taxing. We need to get the attention or focus of the RAS in order to ensure that input is received. Particular actions by a teacher or learner can help to command attention.

For instance, when the learning process (or lessons) become repetitive or predictable then the learner will find it harder to concentrate. By varying presentation styles, activities and resources used by the teacher can introduce an element of surprise which is more likely to engage the learner. The use of the teaching space (through teacher movement around the classroom) can keep the student focused, as can student-led activities introduced at the beginning of a lesson. The RAS is particularly responsive to novelty or change, so using a new display, music, drama, costume, role play can all be received as novel and will alert the RAS to pay attention and promotes the curiosity of the learner (Wang et al. 2005).

External factors can make it more difficult to hold attention. When our glucose levels are low our attention wains. Processing information uses glucose in the brain, and when our glucose drops we become tired and slows down our ability to process cognitive tasks. Similarly, learning in a risky environment is an impediment to focus. Risk can range from a physical risk to emotional risk (like being embarrassed by a teacher or classmates). The RAS gives priority to whatever is presenting as the perceived risk, at the expense of the lesson. Neuroimaging technology shows us that when the threat or risk (evident in the lower brain) is reduced, more sensory input gets through the RAS.

Information must be received and deliberately processed through our working memory into this long-term memory.

Working memory is limited, on average, it is thought that individuals can hold up to seven chunks of information in their working memory at any one time (Miller 1956; Baddeley 1992). In addition, working memory lasts only a short duration – less than one minute. Clark, Kirschner and Sweller call it 'the limited mental space in which we think' (2012: 8). Conversely, long-term memory is unlimited, and is where we make sense of the information that we take in via working memory. Knowledge that is processed and stored in long-term memory is stored using schemata, or frameworks that integrate various elements of information. Schemas are systems for organising knowledge. For example, the squiggles on this page are letters; letters when put together form words, words when put together form sentences. The letters, words and sentences are increasingly complex schemas which allows the reader to process information and make meaning. The more someone reads, the less effort or cognitive load is required.

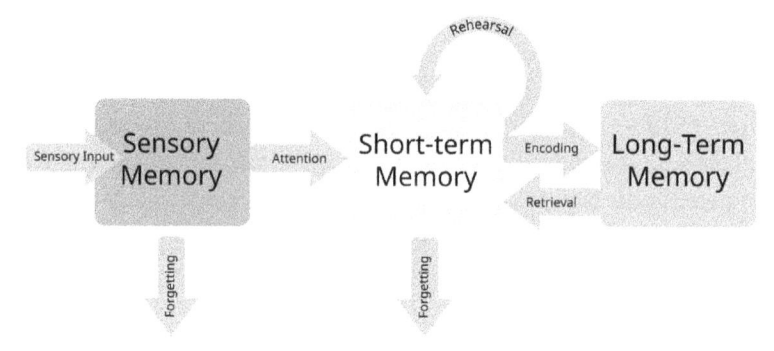

Figure 12: Working Memory – Image by Piscine26/<stock.adobe.com>

It's important when understanding how we learn that we have a certain brain capacity, for if we overwhelm that capacity through cognitive load, we can find a task challenging or give up altogether. If our working memory (processing) is overloaded, there is a risk that the information will not be fully understood or will be confused or misinterpreted which can have a negative impact on learning and long-term memory. If we are distracted or multi-tasking, we can become cognitively overloaded. Imagine trying to read, while also hearing music coming from the next room, and habitually checking phone notifications. The simple task of reading becomes more difficult and requires more effort to understand the text. The more distractions or ineffective load we carry, the more we are susceptible to cognitive overload. Removing distractions can be beneficial, thus managing the complexity of the task can prevent the learner from becoming overwhelmed.

The process of building connections for learning can start within 15 minutes of receiving new information and with rehearsal gains strength as time moves on. The new information must first be imprinted or codified for it to stay within memory. Otherwise the information is lost. Therefore the key for any learner or educator is to ensure that the brain is not so overloaded with new information that there is no time to imprint it as it is received.

Crucial to imprinting is the movement of information from short term to long term memory – work that is done within or by the hippocampus. Part of this process requires a physical recycling of proteins within the neurons of the brain. Learning or imprinting improves when there is manageable content received, followed by a period of rest. We have all experienced a situation where we return after taking a break with a fresh perspective on what has been learned. In fact, numerous researchers have found that much of our memory work occurs while we sleep (Piegneux et al. 2001; Goda and Davis 2003). At night, the new information we receive during the day is codified and organised. Over the following days and weeks, if stimulated, further connections can be made and learning can be solidified. Therefore learning is a complex and lengthy process. This is why cramming is heralded as an ineffective study technique.

Frequently teachers use such strategies as discussion time, reflective journaling and quizzing but these may be more effective if used after the brain has more time to process the information. Furthermore, the heavier the content and the less prior knowledge that the learner has, the more time the brain will take to process and codify the knowledge. Some researchers suggest that this can take as much as 2–5 minutes for every 15 minutes of new information received. During this time the brain is trying to make the content meaningful by relating it to prior information or relating it to the context around the learner.

Scaffolding Learning through Experimental Learning, Emotion and Reinforcement

As we grow, we need to be challenged, we need to make mistakes and we need to be provided with effective learning strategies that can help us to build confidence and competence in learning (Nye et al. 2018). Teaching methodologies that engage learners and command their attention have been found to produce better learning outcomes. According to Radin, a US researcher who undertook a qualitative study of brain-compatible instructional characteristics, brain-compatible teaching is the application of a meaningful group of principles that represents our understanding of how our brain works in the context of education. She found that there are a number of principles that contribute to brain compatible instruction, including creating emotional links with learning; providing safe learning environments; encouraging problem solving, trial and error; critical thinking and problem-solving. In this context, students do the work of learning and create their own meanings (2009: 44). Some of these teaching strategies and the evidence underpinning them are outlined in the following sections.

Spaced Practice and Interleaved Learning

To develop long-term enduring memories, research suggests that we need to distribute learning over time at short intervals. Spaced or distributed practice involves spreading out learning sessions over time, with intervals between each session. Distributed practice involves breaking up study sessions into shorter, spaced intervals over time. For example, instead of studying a subject for 2 hours in one sitting, learners might study for 30 minutes on multiple occasions across several days or weeks. Research consistently demonstrates that distributing learning over time enhances long-term retention compared to massed practice. The spacing effect, a cognitive phenomenon, suggests that information is better remembered when study sessions are spaced out. Distributed practice often involves active recall, where learners actively retrieve information from memory. This retrieval practice reinforces neural connections and enhances the ability to recall information during subsequent study sessions.

Spaced learning (often referred to as distributed learning) is a learning strategy in which two or more study periods are separated in time by an inter-study interval (ISI), the interval can be a brief 10-minute interval or can be months. Studies from cognitive psychology which focused on spacing effects and experiments with spacing and recall, have consistently found that long spaces of a day or more are increasingly beneficial for learning. Moss (1996) reviewed 120 articles on the spaced learning effect, comparing various types of learning material (verbal information, intellectual skill, and motor learning). Longer spacing intervals improved the learning of verbal information and motor skills in over 80 per cent of the studies reviewed. In 2006, Cepeda et al. reviewed 184 articles which included 317 spacing effect experiments with children, adults and older adults. These were all verbal memory learning tasks, followed by a recall test. All but twelve showed a benefit of spaced learning over massed study.

Other studies have identified that spaced learning results in greater recall when learning is repeated. When there is no interval, the brains response has been found to be weaker (Mollison and Curran 2011; Xue et al. 2012). Carpenter et al. (2009) conducted learning sessions with 8th graders in the US. Children were given a history test, then a review of answers and facts. This review was repeated by 1 group 1 week after the test, and by a second group 6 weeks after the test. The test was given again to all children 9 months later. The children who reviewed the answers and facts 1 week after the test recalled less than the children who reviewed 16 weeks after the test. This suggests that the interval period has an important effect on retention and recall.

To learn effectively, and develop deep understanding and knowledge of a topic, we have to have clear meaning and be able to apply it in different contexts. David Perkin's theory of deep understanding focuses on not only possessing knowledge, but also 'being able to think about, explain, and apply it beyond the classroom' (1998: 40). Deep learning develops when a concept is revisited over time, with space between sessions. One method of creating spaced learning or practice, is to use interleaved learning. Interleaving means tackling different problems or ideas in a sequence. That is learning the key concepts of one topic, then moving on to learn the basics of a second topic, then returning to the first topic to revisit and deepen the learning. Returning to a topic regularly forces the learner to recall the skills and knowledge that they need. Spaced learning applies to child and adult learners alike. It is possible that spaced learning, which provides time for the brain to sleep in between periods of learning and recall, may contribute to the success of memory formation (O'Hare et al. 2017).

Expert Insight:

Professor Ian Robertson, Trinity College Dublin

In conversation with the authors, Prof Robertson, Emeritus Professor of Psychology and Co-Director of the Global Brain Health Institute, highlighted some of the key learnings from neuroscience and how they can enhance our understanding of teaching, learning and education more generally.

Prof Robertson said 'evidence suggests that by activating the prefrontal cortex of the brain, by engaging learners in active learning, or active questioning, the brain is put into a mode of trying to create a framework itself rather than accepting a given framework and responding to it. When learners are encouraged to adopt this kind of active mode, deeper learning happens. Just listening to words all the time or even acting on words, is a poor substitute for actually carrying out an experiment or action. Questions which encourage the learner to reflect (metacognition) activates so many more circuits in the brain and the more regions of the brain that are activated, the more secure will the learning be'.

Interleaved learning leads to better consolidation of new information

'Supposing you're trying to learn French, and then do mathematics, and then study history. The evidence is that if you switch, rather than doing big chunks of one topic, if you switch (interleave) different domains of learning, in the shorter term that leads to slower learning that is more effortful. It's harder but it seems to produce more secure long-term retention of your learning. So, if students want their learning to last longer than the day of the exam, they should interleave different domains of learning when studying'.

It is also very important to do nothing, well apparently 'doing nothing', to ensure consolidation of learning. Research and brain imaging on the default mode network, which is in the middle of the brain shows that our brain is very active when we're doing nothing (externally). If you've just learned something, and then you are tested half an hour late and then a week later, those who had less active time after learning retain the information for longer. If you never give the daydreaming network a chance to just do its job (which is actually sorting and storing the information you've learned), then you don't retain information. We need to make sure we give ourselves brain breaks – periods when we just sit there, listening to some relaxing music or going for a walk. To give the filing and storage systems of your brain a chance to work. A brain break does not work if we go on Facebook or play a game.

Confidence influences the brains response to learning

Prof. Roberston spoke about his own research and how it can shed light on the link between education and neuroscience; 'my latest book was on confidence. And confidence is incredibly important. Because confidence is the belief that you can do something. Having that belief actually makes it more likely, you will then do it successfully because of the effects of confidence on the brain. A lot of learners, when learning a musical instrument, gave up because they had maybe not performed well or not progressed. We tend to acquire the belief that you cannot do it. That is almost never accurate for a human being because the brain is so plastic. It's a learning machine, actually. So, we're designed for learning; that is our brains been physically changed by what we learn.

One of the enemies of confidence is having this belief that what you are and what you can do is determined genetically for example, you were born clever, or you were born stupid, or you were born happy or anxious etc. Yes, maybe some inherited an awful lot, but it is also to do with the application of learning, with good teaching or bad teaching, with luck or circumstance. Learners who feel they are destined to fail, risk their brains responding completely differently. In brain imaging, children with a fixed mindset, if they're told they did something wrong, their hippocampus (the memory centre) closes down and the parts of the brain that are related to self-evaluation, light up. That means that the brain goes into a non-learning state'.

Noradrenaline can be controlled and can impact learning

In my own research, I also study arousal and attention. I have found that there is this optimal level of arousal as measured, by the noradrenaline system of the brain, which is part of the fight or flight system. Optimal learning happens when there is just the right amount of noradrenaline. Feelings of boredom, low mood, lack of stimulation, tiredness, lack of sleep etc. give the brain too little noradrenaline, which means it can't pay attention too well, and can't learn so well. The opposite is also true – threats, competition etc. produce too much and the brain can't learn properly. So, there's a sweet spot of arousal and attention. That sweet spot is probably the critical learning indicator. If people can control their brain in such a way as to keep relatively near the sweet spot, their learning and wellbeing will be much improved. That level of arousal can be controlled by your breathing for example, slow breathing can reduce noradrenaline levels to make it easier to perform.

Recent Research

Robertson, I. (2021). *How Confidence Works: The New Science of Self-belief*. London: Transworld Publishers Ltd. 978-1787633728

Robertson, I. (2018). *The Stress Test: How Pressure Can Make You Stronger and Sharper*. London: Bloomsbury. 978-1408860397

Robertson, I. (2013). *The Winner Effect: How Power Affects Your Brain*. London: Bloomsbury. 978-1408831656

Self-regulation and Emotional Connection

Sensory input, once it passes through the reticular activation system (RAS), must be processed by the prefrontal cortex (PFC). The PFC regulates cognitive and executive functions. Positive emotional influences also get priority for memory and dopamine (a neurotransmitter) is linked to the brains reward system, furthermore hormones epinephrine and norepinephrine enhance memory. Therefore, pleasurable learning tasks, and happy learners, are more likely to process stimuli through the PFC and generate associations or connections which support recall. Active processing, through activities like questioning and reflection, allows learners to take charge of the consolidation and internalisation of learning in a way that is personally meaningful to them (Caine and Caine 1989).

There is growing evidence of links at the neural level between emotion and learning, confirming what many teachers will have experienced in the classroom. A range of studies have highlighted the importance of emotions in helping the brain to prioritise information that is eventually brought to the learner's focal awareness (Compton 2003; Vuilleumier 2005). We also know that stress and threat affect the brain, and it reacts differently in relation to experiences such as peace, boredom and contentment (Ornstein and Sobel 1987).

In this regard the functioning of the amygdala is affected by negative emotional influences like stress. Stress triggers hormonal responses which negatively impact on how memories are formed. Those learners who can control and regulate their stress levels, for example during a test, are likely to find it easier to learn, remember and use information. Psychological research suggests that self-control may reflect the formation of productive habits and that this in turn reduces the need for effortful self-control (Galla and Duckworth 2015; Fiorella 2020). A classic example of self-regulation is the marshmallow test outlined in the experiment below.

Furthermore, emotion has a substantial influence on cognitive processing, because it affects, not only what it is the learner focuses on, but also their motivation and behaviour (Tyng et al. 2017). Triggering emotional responses can help learners to direct attention to information from the environment, processing it from the thalamus to the amygdala and on to the prefrontal cortex. Emotional connections or responses could be triggered by encouraging learners to experience the emotions in an event rather than just learning about it. For instance, hearing about personal stories, role playing, discussions, debates, provocations, and practical demonstrations are likely to produce more emotional responses than didactic lessons.

Experiment

The Marshmallow Experiment is a well-known psychological study on delayed gratification conducted by psychologist Walter Mischel in the 1960s at Stanford's Bing Nursery School. The study aimed to investigate how children's ability to delay immediate rewards for a better reward later in life correlated with various life outcomes. The study involved preschool-aged children; 4–6 years old. Children were brought into a room with a table and a single marshmallow placed on it. The researcher would tell each child that they could either eat the one marshmallow immediately or, if they could wait for a short period, they would be rewarded with two marshmallows. The researcher then left the room (for about 15 minutes), leaving the child alone with the marshmallow. He found that about one third of the pre-schoolers ate the marshmallow immediately; a second third were able to wait but not for the full 15 minutes; and, a third that waited the full 15 minutes and received the second marshmallow. Mischel conducted follow-up studies to assess the long-term impact of a child's ability to delay gratification. They found correlations between a child's ability to delay gratification in the Marshmallow Experiment and various life outcomes, such as academic success, health, and social competence. In particular, he more seconds they waited at the age of four or five, the higher their SAT scores (an important college entrance examination in the US) and the better their rated social and cognitive functioning as adolescents (Mischel 1961).

Watch: <https://www.youtube.com/watch?v=jnq2LmanNOYY>

Patterning

In cognitive psychology, patterning refers to the process of recognising patterns in information, events and experiences. The human brain is wired to identify regularities and make sense of the world by organising information into coherent patterns. The search for meaning occurs through patterning: 'The brain is designed to perceive and generate patterns, and it resists having meaningless patterns imposed on it' (Caine and Caine 1994: 88). This cognitive patterning helps individuals derive meaning from their surroundings and experiences. In language acquisition, patterning can look like recognising grammatical patterns and structures. Learners identify recurring patterns in sentence construction, vocabulary usage, and language rules, which support language proficiency. In a social context, learners may recognise patterns of cultural norms, values and narratives which can help them to derive meaning from their relationships. Learning is also built on the process of detecting and making patterns. Emotions are critical to patterning: in the ground breaking *The Emotional Brain*, Joseph LeDoux (1996) clearly explains how the emotional neural passageways (which originate in our amygdala) influence the neural passageways needed for academic and scholarly work.

Consolidating Learning through Rehearsal, Recall and Transfer

Consolidation of learning refers to the process by which newly acquired information is strengthened, and integrated into existing knowledge. It is a crucial step in the learning process that helps to move information from short-term memory to long-term memory, making it more durable and accessible over time. Consolidation facilitates this transfer and is supported through a number of steps, including rehearsal or practice, recall or retrieval and transfer of learning to different contexts.

Consolidation through active rehearsal and practice helps store our knowledge in long term memory and frees up space in our working memory. Consolidation involves changes at the synaptic level, where connections between neurons are strengthened. This may include the formation of new synapses or the modification of existing ones. These structural changes contribute to the lasting storage of information. According to Howard-Jones (2020), research with learners has found that on brain scans, when learners are engaging with new information, the frontal working memory networks light up. Once the information is consolidated and stored in long term memory, activity is reduced in the frontal networks, and is instead evident in regions in the back of the brain.

Rehearsal

Rehearsal and practice play a significant role in the consolidation of learning. Neuroscience is beginning to make sense of what is happening in the brain when we make memories – and for scientists, everything that is learned is memory. Part of this process of storing information is called encoding. This is the process whereby new knowledge is converted to a form that can be stored. Our brains are multi-sensory and information on a given subject can be encoded by visual, auditory or sensory means. These means can be interconnected. For example, an image of a cat may trigger encoding of visual, auditory and sensory means. These complex interactions and connections between stored information and knowledge systems in the brain help us to interpret our daily reality.

Rehearsal is the cognitive process of repeating or reviewing information as a way to enhance learning and memory. When we rehearse knowledge, various cognitive processes and neural mechanisms come into play, contributing to the encoding, consolidation, and retrieval of information in the brain. Repeated exposure to knowledge through study or practice helps strengthen the neural connections associated with that knowledge. As we rehearse information, the hippocampus helps in organising and initially storing the knowledge in short-term memory. With continued rehearsal and practice, there is a gradual transfer of information from short-term to long-term memory. The hippocampus works in conjunction with the neocortex which serves as a long-term storage site for memories. Rehearsal induces synaptic plasticity, which refers to the ability of synapses (junctions between neurons) to strengthen or weaken over time.

Neuroscience supports the idea that learning is about building connections or associations between existing ideas to form new concepts. Our comprehension of the world around us is dependent on our prior knowledge. Lee Shulman in his book *The Wisdom of Practice* writes, 'to prompt learning, you've got to begin with the process of going from inside out'. Rehearsing knowledge in different ways will connect it with prior knowledge. Connections between new concepts and prior knowledge will strengthen and consolidate the learning. Repeated rehearsal establishes associative connections between the new information and existing knowledge. These associations contribute to a more interconnected and organized knowledge network in the brain. The more connections that are made in the brain, the more ways there are to trigger recall. Encouraging students to enact, discuss and express their new knowledge will help them to make more connections in the brain.

When learners have appropriate, retrievable, and accurate prior knowledge to link new knowledge to, learning will happen more readily and will endure. Emotional engagement during rehearsal can also enhance the consolidation process. Information with emotional significance is often better remembered, and the amygdala's involvement can influence the consolidation of emotionally charged memories. Some of the key techniques which are regularly used to support rehearsal include having learners review information on flashcards and actively recalling answers; having learners repeat key facts, definitions, or concepts multiple time; asking learners to summarise or synthesis information in their own words; providing opportunities for learners to teach others to reinforce understanding; and creating visual cues like mind maps which provide representations of the relationships between concepts.

Recall

Recall or retrieval is the process of accessing and producing information that has been previously learned and stored. Recall can be both intentional, where we actively try to remember something, or spontaneous, where information comes to mind without deliberate effort. When we learn new information, the ability to retrieve that information during assessments, discussions, or practical applications demonstrates that the learning has been successful. In the brain, retrieval practice can be described as the process of transferring 'information to mind from memory' (Weinstein et al. 2019: 85). Retrieval cue comes from sources like a question in a book or even a teacher's verbal question. Those prompts or cues are what the learner then uses to search their memory. Learners can also create their own cues using study notes or mind maps.

When revising for an exam or assessment, the revision strategies used are essential to ensure successful recall. According to Dunlosky (2013), learners can fall into the trap of thinking that conventional strategies are effective, because they have not been taught how to go about learning the content and what skills will promote efficient studying. Dunlosky argues that practice testing and distributed practice are most effective because 'they can help students regardless of age, they can enhance learning and comprehension of a large range of materials, and, most important, they can boost student achievement' (p. 13). Practice tests allow students to identify what they know and what they do not know, thus triggering a need for further study. Similarly distributed practice requires students to practise consistently rather than cram, which will result in a stronger ability to recall over time. According to William (2002), the theory of distributed practice or the spacing effect has held up well over more than one hundred years of research, and has been found to benefit both school-aged children and older learners (Cornelius Rea and Vito Modigliani 1985).

Transfer of Learning

Transfer of learning is the ability to extend what has been learned in one context (the classroom) to another context (practice). Brain research tells us that each time we recall an event or a previous experience, our brain reconstructs the experience through the same circuit or circuits where we used to store it. Teaching for transfer is possible, but it is much more easily stated than accomplished (Craig 2003: 346). By incorporating certain strategies into teaching, teachers and learners can contribute to a learning environment that promotes the transfer of learning. For example, teachers and learners can enhance their ability to transfer by using case studies and examples, by encouraging deep understanding of concepts rather than rote learning; by connecting classroom learning to real-world situations; by emphasising the development of transferable skills, such as critical thinking, problem-solving, and effective communication; by drawing parallels between familiar ideas and new concepts can aid in understanding and promote the transfer of learning, and by designing assignments and assessments that require the application of knowledge in novel situations.

> **Podcast**
>
> This episode of the Podcast Talking Teaching and Learning, Kevin L. Merry speaks to Burt Oraison from Victoria University in Melbourne discusses Deep Learning on the block. Bert discusses the ways in which we can approach Block teaching session design and delivery to better encourage learners to learn deeply. This is a must listen episode for anyone interested in effective learning, teaching and assessment practices.
>
> Available from: <https://podcasters.spotify.com/pod/show/kevin-merry/episodes/Block-Series-Episode-3-Deep-Learning-on-the-Block-with-Bert-Oraison-e24ti1q/a-a9ttocl>

> **Resource**
>
> Supporting Learner Revision with the *Seven-step Model*. Julie Kettlewell from the Education Endowment Foundation created two planning tools for teachers, informed by metacognition and self-regulated learning research, to help scaffold pupils' revision. The resources can be found here: <https://educationendowmentfoundation.org.uk/news/supporting-revision-and-the-seven-step-model>

Metacognition

Influenced by the developmental psychology of Piaget, psychologist John Flavell introduced the term 'metacognition' in the 1970s as a result of his research with children on control of the memory processes. Flavell (1976) defined metacognition as the active monitoring and consequent regulation and orchestration of a variety of information processing activities. In 1987, he described a variety of examples such as feeling that one is not understanding something, feeling that something is difficult or easy to remember, solve, or comprehend, and feeling that one is approaching or failing to approach a cognitive goal. Metacognition is more than simply thinking about thinking; it is also reflecting upon one's own thinking and altering behaviour based on learning.

Teachers can scaffold the development of learners' metacognitive skills by helping learners to monitor their own progress and take control of their learning as they read, write and solve problems in the classroom. According to Darling-Hammond et al., 'To enable students to manage their own learning and transfer it to new contexts, teaching should be designed to develop students' *metacognitive capacity, agency, and the capacity for strategic learning* (2019: 111).

Metacognition amongst learners is encouraged by creating a classroom environment where students' mistakes are accepted, and where teachers highlight their own mistakes as examples of learning moments.

Furthermore, allowing time for discussion, providing opportunities for self-direction, asking reflective questions and incorporating elements of self-assessment and peer assessment can encourage metacognition amongst learners.

Significant work has been done on integrating metacognitive practices in both reading and maths instruction. Readers have been encouraged to monitor their understanding in the process of reading (Pearson, Cervetti, and Tilson, 2008). Better readers have better metacognitive awareness of their own strategy use, which leads to enhanced reading ability (Hamdan et al. 2010); in maths instruction, verbalising and writing the steps to solving a problem help students reflect on, monitor, and evaluate their problem-solving abilities and strategies. This has been shown to increase conceptual understanding and provides students the opportunity to evaluate their learning (Martin, Polly, and Kissel 2017). That said, metacognitive skills are applicable to all subjects of the curriculum, and can be transferred from subject-to-subject.

'Metacognition can be more broadly understood as a part of self-regulated learning (SRL) and connected with the emotional and motivational dimensions of learning students take responsibility for their own learning and are active in the learning process' (Muijs and Bokhove 2020). The learner must (a) be able to make an internal judgement about one's knowledge or cognitive process and (b) the learner must be able to control or self-regulate their cognitive processes by adapting to the outcome. Giving students the time and space to reflect on their own thinking is critical for fostering student agency.

Blog Post

In her blog post, *Showing Students How Metacognition Works*, Tricia Taylor explains that metacognition is simply good learners thinking about their learning, which helps them to make good decisions and learn from their mistakes. She argues that we can teach metacognitive regulation to students through explicit instruction, lots of modelling and practice (Hattie 2011; Hattie et al. 1996; Dignath and Buttner

2008). The key for teachers is to be explicit about the strategies they are using, model the practice, relate the knowledge to a context or subject, incorporate verbalisation (think, pair share or similar), and allow time for practice. In this blog, she advocates for the Model, Teach, Practise, and Connect step-by-step guide to teach metacognition.

Available from: <https://www.learningscientists.org/blog/2021/6/10-1>

Website

According to Lee Davis Deputy Director of Cambridge International Education, metacognition is a process whereby students plan, monitor and evaluate their thinking around a particular learning objective. It's about understanding what success will look like and identifying strategies that will help them reach their goal. 'Too often we teach students *what* to think but not *how* to think' (OECD 2014). The Cambridge International Teaching and Learning Team have prepared resources which investigate what metacognition is, what the research tells us and how we can benefit from metacognition. They include a list of resources, books, articles and websites that support teachers and learners to understand and practice metacognition in the classroom. Some key questions for teachers include:

Have I included clear learning objectives?
How am I going to encourage my students to monitor their learning?
How can I create opportunities for learners to practice new strategies?
How can I allow time for learner self-reflection?
Does the classroom environment support metacognitive practices?

Available from: <https://cambridge-community.org.uk/professional-development/gswmeta/index.html>

Conclusion

Learning happens as the brain processes information by making and breaking connections, growing and strengthening the synapses that connect neurons to their neighbours. But once a potential network is in place, it needs to be used if it is going to continue functioning. Understanding how brain and mind functions interact is a huge challenge for those interested in teaching and learning. Even if we can use neuroscience and neuroimaging technology to pinpoint what is happening in the brain at the time of learning, the definition of learning itself is about the cognitive process of understanding, metacognition and recall.

We argue that it is important to continue to learn throughout our lives, and to be open to new modes of learning. As the United Nations Educational, Scientific and Cultural Organisation (UNESCO) argued in *Learning to be* (Faure et al. 1972), the concept of lifelong learning includes the crucial importance of learning throughout life for enriching personal lives, fostering economic growth, social inclusion and cohesion. The notion of lifelong learning has also opened up our thinking of learning as something broader than what goes on in school. Non-formal, informal and self-directed learning are legitimate sites for learning, and this broader perspective of learning is gradually transforming teaching practice. Insights from neuroscience research will continue to bring additional knowledge and understandings for learners in all sites for learning.

References

Atkinson, Richard C. and Shiffrin, Richard M. (1971). The Control of Short-Term Memory. *Scientific American*, 225(2), 82–91.
Baddeley, A. (1992). Working Memory. *Science*, 255(5044), 556–559.
Bandura, A. (1965). Behavioural Modification through Modelling Procedures. In L. Krasner and L. P. Ullman (Eds), *Research in Behaviour Modification*. New York, NY: Holt, Rinehart & Winston.
Bandura, A. (1977). *Social Learning Theory*. Englewood Cliffs, NJ: Prentice Hall.
Bandura, A. (1986). *Social Foundations of Thought and Action: A Social Cognitive Theory*. Prentice-Hall.
Bronfenbrenner, U. (1992). Ecological Systems Theory. In R. Vasta (Ed.), *Six Theories of Child Development: Revised Formulations and Current Issues* (pp. 187–249). London: Jessica Kingsley Publishers.

Caine, G. and Caine, R. N. (May 1989). Learning about Accelerated Learning. *Training and Development Journal*, Vol 43(3) 65–73.

Carpenter, S. K., Pashler, H. and Cepeda, N. J. (2009). Using Tests to Enhance 8th Grade Students' Retention of US History Facts. *Applied Cognitive Psychology*, 23, 760–771.

Censor, N., Sagi, D. and Cohen, L. (2012). Common Mechanisms of Human Perception and Motor Learning. *National Review of Neuroscience*, 13(9), 658–664.

Cepeda, N. J., Pashler, H., Vul, E., Wixted, J. T. and Rohrer, D. (2006). Distributed Practice in Verbal Recall Tasks: A Review and Quantitative Synthesis. *Psychological Bulletin*, 132, 354–380.

Chandler, P. and Sweller, J. (1991). Cognitive Load Theory and the Format of Instruction. *Cognition and Instruction*, 8(4), 293–332.

Committee on How People Learn II: The Science and Practice of Learning. (2018). *How People Learn II: Learners, Contexts and Cultures. National Academies of Sciences, Engineering and Medicine*. Washington, DC: The National Academies Press. Available from: <https://www.informalscience.org/sites/default/files/how%20people%20learn%20ii.pdf>

Compton, R. J. (2003). The Interface between Emotion and Attention: A Review of Evidence from Psychology and Neuroscience. *Behavioral and Cognitive Neuroscience Reviews*, 2(2), 115–129.

Darling-Hammond, L., Flook, L., Cook-Harvey, C., Barron, B. and Osher, D. (2019). Implications for Educational Practice of the Science of Learning and Development. *Applied Developmental Science*, 24(2), 97–140.

Dehaene, S., Pegado, F., Braga, L. W., Ventura, P., Nunes Filho, G., Jobert, A., Dehaene-Lambertz, G., Kolinsky, R., Morais, J. and Cohen, L. (2010). How Learning to Read Changes the Cortical Networks for Vision and Language. *Science*, 3;330(6009), 1359–1364.

Elliott, L., Gascoine, L., Fairhurst, C., Mitchell, A., Fountain, I., Bell, K. and Torgerson, D. (2019). *Sc-inapse: Uncertain Rewards*. Education Endowment Foundation. University of Bristol.

Faure, F., Herrera, F., Kaddoura, A., Lopes, H., Petrovsky, A., Rahnema, M. and Champion Ward, F. (1972). *Learning to Be: The World of Education Today and Tomorrow*. Paris: United Nations Educational, Scientific and Cultural Organisation (UNESCO).

Fiorella, L. (2020). The Science of Habit and Its Implications for Student Learning and Well-being. *Educational Psychology Review*, 32, 603–625.

Friston, K. (2002). Learning and Inference in the brain. *Neural Networks*, 16(1), 1325–1352.

Galla, B. M. and Duckworth, A. L. (2015). More Than Resisting Temptation: Beneficial Habits Mediate the Relationship between Self-control and Positive Life Outcomes. *Journal of Personality and Social Psychology*, 109(3), 508–525.

Gazzaniga, Michael S. (Ed.). (2009). *The Cognitive Neurosciences* (4th edn). Boston: Massachusetts Institute of Technology. Available from: <https://www.hse.ru/data/2011/06/28/1216307711/Gazzaniga.%20The%20Cognitive%20Neurosciences.pdf>

Goda, Y. and Davis, G. W. (2003). Mechanisms of Synapse Assembly and Disassembly. *Neuron*, 40(2), 243–264.

Hebb, D. O. (1949). *The Organization of Behavior: A Neuropsychological Theory*. New York, NY: John Wiley.

Howard-Jones, P. (2014). Neuroscience and Education: Myths and Messages. *Nature Review Neuroscience*, 15, 817–824.

Howard-Jones, P. A., Varma, S., Ansari, D., Butterworth, B., De Smedt, B., Goswami, U., Laurillard, D. and Thomas, M. S. C. (2016). The Principles and Practices of Educational Neuroscience: Comment on Bowers (2016). *Psychological Review*, 123(5), 620–627.

Kitayama, S. and Salvador, C. E. (2017). Culture Embrained: Going Beyond the Nature-Nurture Dichotomy. *Perspectives on Psychological Science*, 12(5), 841–854.

Kullberg, A., Runesson Kempe, U. and Marton, F. (2017). What Is Made Possible to Learn When Using the Variation Theory of Learning in Teaching Mathematics?' *ZDM Mathematics Education*, 49, 559–569.

Mareschal, D., Butterworth, B. and Tolmie, A. (Eds). (2013). *Educational Neuroscience*. Massachusetts, USA: Wiley and Sons. Available from: <https://perpustakaan.gunungsitolikota.go.id/uploaded_files/temporary/DigitalCollection/M2I4MGM5ZmVkODZmMjRiOTgwOWY3M2Q2MmUyYWExMzYyN2U2ZTBmZg==.pdf>

Martin, C. S., Polly, D., & Kissel, B. (2016). Exploring the impact of written reflections on learning in the elementary mathematics classroom. *The Journal of Educational Research*, 110(5), 538–553.

Miller, G. A. (1956). The Magical Number Seven, Plus or Minus Two: Some Limits on Our Capacity for Processing Information. *Psychological Review*, 63, 81–97.

Mischel, W. (1961). Delay of Gratification, Need for Achievement, and Acquiescence in Another Culture. *Journal of Abnormal and Social Psychology*, 62, 543–552.

Mischel, W. (1981). Metacognition and the Rules of Delay. In J. H. Flavell and L. Ross (Eds), *Social Cognitive Development: Frontiers and Possible Futures*. New York: Cambridge University Press.

Mollison, M. V. and Curran, T. (2012). Familiarity in Source Memory. *Neuropsychologia*, 50, 2546–2565.

O'Hare, L., Stark, P., McGuiness, C., Biggart, A. and Thurston, A. (2017). *Spaced Learning: The Design, Feasiblity and Optimisation of SMART Spaces.* Evaluation Report.Education Endownment Foundation.

Ornstein, R. and Sobel, D. (1987). *The Healing Brain: A Radical New Approach to Health Care.* Oxfordshire: Macmillan.

Pearson, P. D., Cervetti, G. N., Tilson, J. L. (2008). Reading for understanding. In L. Darling-Hammond, B. Barron, P. D. Pearson, A. L. Schoenfeld, E. K. Stage, T. D., Zimmerman, G. N. Cervetti, & T. L. Tilson (Eds.), Powerful learning. What we know about teaching for understanding (pp. 71-112). San Francisco, CA: John Wiley & Sons.

Perkins, D. (1998). What Is Understanding? In M. S. Wiske (Ed.), *Teaching for Understanding: Linking Research with Practice.* San Francisco, CA: Jossey-Bass.

Piegneux, P., Laureys, S., Delbeuck, X. and Maquet, P. (2001). Sleeping Brain, Learning Brain: The Role of Sleep for Memory Systems. *Neuroreport*, 12(A111–A124).

Radin, J. L. (2005). *Brain Research and Classroom Practice: Bridging the Gap between Theorists and Practitioners.* Doctoral dissertation, Colorado State University.

Reif, F. (2010). *Applying Cognitive Science to Education: Thinking and Learning in Scientific and Other Complex Domains.* Cambridge, MA: The MIT Press.

Schapiro, A.C. and Turk-Browne, N. B. (2015). Statistical Learning. In Arthur W. Toga (Ed.), *Brain Mapping: An Encyclopaedic Reference* (pp. 501–506). Academic Press/Elsevier.

Simmonds, A. (2014). *How Neuroscience Is Affecting Education: Report of Teacher and Parent Survey.* Wellcome Trust. Available from: <https://wellcome.org/sites/default/files/wtp055240.pdf>

Tyng, C. M., Amin, H. U., Saad, M. N. M. and Malik, A. S. (2017). The Influences of Emotion on Learning and Memory. *Frontiers in Psychology*, 8.

Vuilleumier, P. (2005). How Brains Beware: Neural Mechanisms of Emotional Attention. *Trends in Cognitive Sciences*, 9, 585–594.

Wang, B., Shaham, Y., Zitzman, D., Azari, S., Wise, R. A. and You, Z. B. (2005). Cocaine Experience Establishes Control of Midbrain Glutamate and Dopamine by Corticotropin-releasing Factor: A Role in Stress-induced Relapse to Drug Seeking. *The Journal of Neuroscience: The Official Journal of the Society for Neuroscience*, 25, 5389–5396.

Wiliam, D (2017). I've Come to the Conclusion Sweller's Cognitive Load Theory Is the Single Most Important Thing for Teachers to Know. <http://bit.ly/2kouLOq>, tweet,>.

Woloshyn, V. E., Pressley, M. and Schneider, W. (1992). Elaborative-Interrogation and Prior-Knowledge Effects on Learning of Facts. *Journal of Educational Psychology*, 84(1), 115–124.

Wood, W., Quinn, J. M. and Kashy, D. A. (2002). Habits in Everyday Life: Thought, Emotion, and Action. *Journal of Personality and Social Psychology*, 83(6), 1281–1297.

CHAPTER 7

Harnessing Habits for Learning

> 'Habits are like a cable; we weave a strand every day, and soon it cannot be broken'.
>
> (Mann, H. 1850 cited in Covey 2000: 25)

A habit is a well learned response that is carried out when the appropriate stimulus is present. Habits are wired so deeply in our brains that we perform them automatically without effort or thought. They are formed because the human brain is highly efficient. Once a specific habit is formed, it is like a programme that runs on autopilot. When performed often enough, the behaviour becomes difficult to break. Habits include the entire gamut of automatic and goal-oriented behaviours we are engaged in every day, both healthy and unhealthy. They play a powerful role, not only in learning, but more importantly in how we approach formal learning.

Behaviour requires many repetitions before the brain learns to perform a particular function. With the help of the basal ganglia, even the most challenging of routines can become automatic with frequent and consistent action. Most habits start off as positive, goal-directed behaviour – this can be a healthy behaviour like time management or an unhealthy one like procrastination. Repeated often enough, it becomes automatic.

Our habits reflect how we choose to spend time interacting with the world, guided by our short-term and long-term goals. Habits are thought to be driven by reward-seeking mechanisms that are built into the brain. While there is some truth to this thinking as reward mechanisms help habits to form, from the brain's point of view, habits help with efficiency and energy conservation. Our brain turns daily actions and behaviours into habits, so we enact them unconsciously and without much thought, thus freeing up brainpower for other more important challenges. When walking we do not use our conscious mind to determine the motion of our feet, the bend of the knee or the angle of the heel and so on. This is an example of habitual (automatic) movement. This frees our conscious mind to focus on goal-orientated movement or allows us to focus on higher value challenges.

The first challenge in understanding habits is to examine a few pertinent questions. What are habits, and what part of the brain controls them? How are habits formed? Why are they difficult to break? Are our habits beneficial to our brain's capacity to learn, and if not, how can we change them?

What Are Habits?

As humans, we are creatures of habit. William James, the founder of American psychology, popularised the idea of consciousness as a stream of mental events. He realised that, despite our feeling of being a conscious and self-managing our own decisions, we are instead a collection of memories and habits. Most of the choices we make every day may feel like the outcome of deliberate decisions, but they are not. They are habits. According to James, 'our life is but a mass of habits' (1899: xv). It has been estimated that more than 40 per cent of the things one does every day requires no conscious thought whatsoever. According to neuroscientist, David Eagleman, 'brains are in the business of gathering information and steering behaviour appropriately. It does not matter whether consciousness is involved in the decision-making, but most of the time, it is not' (2011: 6).

To take a personal example, when you got out of bed this morning what was the first thing you did? Did you check your phone? Did you make breakfast before you had a shower or after a shower? What route did you take to work? When you got there, what routines did you perform? For us, thinking is expensive. It requires brain activity which takes energy. More importantly, thinking takes time and attention which are limited resources. The more you can safely file away an established behaviour, the more you can concentrate on your current goals and interests.

In *Thinking, Fast and Slow*, Kahneman (2011) identifies two modes of thought which he called System 1 and System 2. System 1 thinking is fast, automatic, frequent, stereotypical, intuitive and unconscious. For example, we can do simple arithmetic, recognise the source of a sound or read an obvious facial expression. In contrast System 2 thinking is slow, effortful, logical, infrequent, calculating and conscious. Here examples include doing more complicated arithmetic, preparing a presentation or sustaining a faster than normal walking rate. He found that for the most part, we make decisions based on System 1. The brain he said can be lazy, preferring to fall back on its well-known strategies and assumptions, and that often leads to errors in accuracy.

It is interesting to apply the fast and slow brain systems to the differences between learning and studying. Learning is a basic instinct and our brain is continually learning; it happens through observation, imitation and everyday events. It is informal and experiential; exciting, enjoyable, unconscious and sometimes emotionally difficult. Learning 'structures' our brain and gives us our personality. Furthermore, this learning occurs without a teacher. It is a constant 'spiral' of growth – taking action, thinking about what we did, making connections with what we already know and bringing it to the next new experience. On the other hand, studying is not natural; it can be hard work and sometimes boring and intellectually challenging. It may induce emotional stress. The act of studying is a formal, deliberate 'habit' of acquiring particular knowledge and it can quickly exhaust our motivation. It cannot be done quickly and requires motivation, repetition, reading, self-testing and writing assignments. A notable feature of studying requires motivation (the will) and cognitive effort (the way), namely, attention, task-switching, working memory and inhibitory control – collectively known as executive function.

It is not just individual habits that become automatic. All organisations have policies, procedures and practices for employees, and these contribute to what is called the culture of the organisation. For example, who sits where in the staff canteen? Is it the norm to be available by phone 24/7? There are many different expressions used to describe organisational culture but essentially the organisational culture generates habitual behaviour. One of the first things a new student has to adapt to in the academic culture are the conventions of being a student, namely the rules and procedures associated with attending lectures, note taking, or finding a book in the library.

Like so many things in life, habits are a double-edged sword. While they can make everyday life easier, they can also become difficult to break. Our brains do not distinguish between healthy and unhealthy habits. While most habits work to save energy, not all our habits are beneficial. As habits ignite reward behaviours, when they are repeated often enough they can become addictions. Unhealthy addictions are a good example of how habits become anchored in the brain relatively quickly but always in an incremental way. Many people come to define themselves by such self-defined negative habits or failures. Sometimes the most difficult habits to change involve the brain's patterns of thinking or relating to other people. As we will see, to understand how to change a habit the first step is to know something about how habits are formed in the first instance.

How Are Habits Formed?

How does the brain form a habit and why? The answer is habituation which is a characteristic of neurons and networks of neurons firing less frequently if we repeat the same action over and over. For instance, in the 1990s, a group of scientists from the Massachusetts Institute of Technology (MIT) discovered a neurological process that is at the core of every habit. They called it the habit formation loop based on evidence from tracking the brain activity of a rat as it ran through a maze (Graybiel 1998).

The researchers trained a rat to learn a new habit by running through a simple T-shaped maze. Over twelve weeks, through constant repetition, the rat learned to either turn left or right depending on a cue (audible tone) which led to a reward (chocolate). Once they had demonstrated that the habit was fully ingrained in the rat's brain, the researchers broke the acquired habit by switching to a new cue. Again, the rat learned a new habit because there was a different reward. However, the researchers were in for a surprise because when they stopped providing the new reward for the second habit, the rat immediately reverted to the initial habit.

What is interesting about this study is that the researchers were able to track what was happening in the brain of the rat as it learned a habit, broke a habit, and reverted back to the initial habit. Neuroscientists already knew that habit-making decisions are formed in the part of the brain known as the basal ganglia. The basal ganglia are a set of large structures towards the centre of the brain that surround the limbic system. They are involved with integrating feelings, thoughts and fine-tuning movement, along with having an important role in forming habits. This fine-tuning is linked to the brain's prefrontal cortex (where thought and planning occurs), which regulates our moment-to-moment control of our habits. This is an important insight, as the brain does not relinquish control of habitual behaviour. Rather, as we will see when talking about forming new learning habits, we can take control of the mechanism by which we make decisions in the prefrontal cortex of our brain.

To return to the MIT experiment, when the rat is first introduced to the maze, neurons in the basal ganglia spike and become active as it picks up cues to locate the reward (in this case chocolate). Initially everything is new, different and important to understand. After several runs, less neural activity is required – it is as if the rat's brain only wanted to focus on what was important (getting to the chocolate) and it no longer needed to take in and analyse everything in its surroundings. The brain was on autopilot so to speak. Because of this efficiency, a new pattern emerged in the rat's neuron activity. This spiked at the beginning when the rat started the maze, then went into sleep mode, and spiked again at the end when the rat licked up the chocolate.

Think back to a time when we first started to perform a new habit. Initially it required a lot of our concentration and brain power. But as we repeatedly went through the routine, it became easier. The mental power we need to perform the task over time decreases significantly. In sport science, this has become known as the groove theory of habits. The theory, first developed in the 1970s, is well presented in Tim Gallway's approach to tennis coaching in *The Inner Game of Tennis*. The theory is a simple one: every time we perform a new behaviour in a certain way, we increase the probability that we will do the same again. In this way, patterns, called grooves, build-up which have a predisposition to repeat themselves. It is as if the nervous system were like a record disk. Every time an action is performed, a slight impression is made in the microscopic cells of the brain, just as a leaf blowing over a fine-grained beach of sand will leave its faint trace. When the same action is repeated, the groove is made slightly deeper. After many similar actions, there is a more recognisable groove into which the needle of behaviour seems to fall automatically.

This grooving process is particularly relevant to learning and studying in higher education. Taking the example of spacing study, this idea implies that the brain retains information when new content is studied in short bursts, and at regular intervals.

Chunking as the Root of Habits

Our routines are usually a group of behaviours rather than a single task, known as *chunking* and are the root of habits. Every day we rely on these 'chunks' of behaviour to carry out our varied daily tasks. Chunking is thus the process of grouping bits of information into a larger unit. For example, consider preparing a dinner. Initially in performing this activity the brain had to consciously focus on each stage of the task. Over time with practice these stages merge into one as the activity became automatic. Chunking in education is a strategy of breaking down information into smaller manageable segments or units, to help students process and remember them. Chunking reduces cognitive load and increases the amount of information that must be held in short-term memory by increasing the amount of information per chunk.

Chunking (also sometimes termed task-bracketing) is particularly important for initiating a routine. In the brain it serves to package up a behaviour considered valuable and worth keeping in its repertoire, so to speak. In his benchmark paper, *The Magic Number Seven, Plus or Minus Two* (1956), George Miller found that immediate memory span (short-term memory) seemed to be limited to the number of chunks of information which can be retained. The implication for students is that the limited capacity of short-term memory can be increased through the chunking process, therefore information will be more easily retained. The brain naturally creates categories to help us remember information, for example birthdays 26111998 (26/11/1998); mobile numbers 0854414589 (085-4414-589); or something as simple as using the cut and paste commands on a computer.

If we consider how chunking works when writing an essay, then its importance becomes obvious. The purpose of writing an essay in higher education is not to tell the reader everything we know about a topic. Rather it is to demonstrate our capacity for critical thinking. Critical thinking refers to the process of analysing, assessing, synthesising, evaluation and reflecting on information, as gathered from observation, experience or reading. It is thinking in a clear, logical, and reflective manner to solve problems or make decisions. Basically, critical thinking is taking a hard look at something to understand what it really means. Chunking is an approach to foster a mind-set of questioning information, ideas and conclusions.

There are several steps in the chunking process: (1) break large amounts of information into smaller units, (2) identify similarities and patterns and (3) organise the information into meaningful units. The above steps are evidence in the construction and writing of an academic essay. In approaching this task, a student would probably ask question such as: What information supports that? (2) How was the information obtained? (3) Who obtained the information? (4) How do you know the information is valid? (5) Are there other possibilities? In correcting essays, academics value the importance of critical thinking in the development of an independent mind.

The Three-Step Habit Loop

Neuroscientists have been able to unlock what changes in the brain when a conscious action turns into a habit. The brain is always in a general state of awareness and balance, even when one is asleep. There is never an actual break in the stream of brain activity. Only when something is novel, difficult or in some other way unusual is it jolted into paying attention to thoughtful processing. Now the brain must adapt to a new set of circumstances by going through what is called the three-step habit loop. The three steps in habit formation are a cue, a routine and a reward.

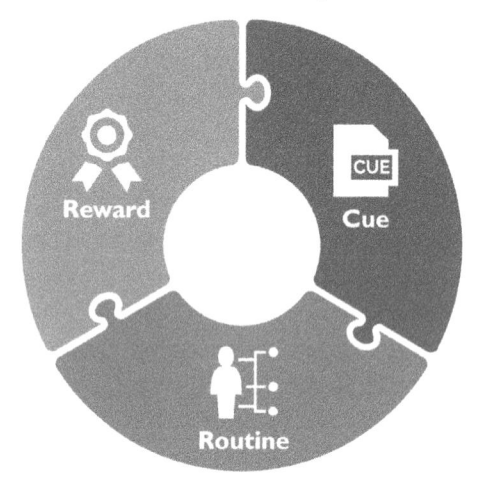

Figure 13: The Habit Loop – Image by Skyline Graphics/<stock.adobe.com>

A Cue

A cue triggers an action that simultaneously creates a desire, longing or craving for a reward. There are different categories of cues that trigger behaviour, such as the immediately preceding action, the location, the time and the emotional state of the individual. Cues differ from person to person. In theory, any piece of information could trigger a cue but in practice people are not generally motivated by the same cues. Cues are meaningless until they are interpreted. It is the thought that transforms a cue into a routine.

A Routine

In the three-step habit loop, the routine is the actual action one takes, which can take the form of an activity, emotion or behaviour. The behaviour may start out as a ritual following a regular procedure but when repeated often enough becomes an automatic routine. When a routine occurs, there is always an element of motivation involved, usually driven by the anticipation of a reward. Routines can be simple or more complex hierarchical systems from motor routines acquired in childhood to the use of language.

A Reward

Our brain is a reward detector. As we go about our lives, our sensory nervous system is continually monitoring what actions deliver pleasure. Psychologists continually remind us that, in the main, pleasurable behaviours are rewarded and reinforced through repetitive learning. This allows the formation of neural pathways in the brain, like grooves that get deeper with use. The neurotransmitter with the strongest capacity to influence the reward system in the brain is dopamine. By supplying our brain with a hit of dopamine from practising a habit, we create a powerful connection between the habit we want to encourage and the body's natural cravings. Think of dopamine as the *save button* in the brain, because when dopamine is paired with a habit, the neural pathways associated with that new habit are solidified and retained.

As noted above, the cue is about noticing the reward. Craving is wanting it while the routine is about obtaining it. Rewards, being the end goal of any habit, close the feedback loop and complete the habit cycle. The reward helps neural 'grooves' to form in the brain. As the action is repeated over and over again, the neural pathway bulks up and the habit becomes entrenched. The more often we perform an action or behave in a certain way, the more it gets physically wired into our brain.

How to Change a Habit

To change a habit, we need to keep the cue and the reward, but change the routine (behaviour). Trying to modify a long-established habit is no small task. Researchers suggest that the first step is to isolate the routine, the behaviour we want to stop. Being aware of the habit and what is reinforcing it is a positive way forward and is the first step to change. Take our response to stress as an example. A typical procrastinator is somebody who copes with stress by putting things on the long finger. While this may serve the purpose of relaxing the person in the short term, in the long term it can become an unhealthy habit which leads to anxiety, further stress, and a lack of productivity and a deep feeling of dissatisfaction.

As we expose ourselves to a displeasing task, our brain relates it to something painful and we immediately turn our attention to a more engaging activity. This gives us a temporary reward in that it feels good because we do not focus on the task as initially planned and instead indulge in a more enjoyable and easier behaviour. There are no limits to our ordinary mundane routines and chores to distract us from studying – hanging out the clothes, checking something on google, responding to e-mails, reading an online article with a catchy heading and so on.

The second step is to recognise the cue or what compels the behaviour. The cue is the trigger that powers the routine and sends a signal to your brain to set off a specific behaviour. Continuing with our procrastination case, the cue is the task you do not want to work on. The brain retains a memory of the habit context, wherein this pattern can be triggered if the right habit cues come back.

The third step is to analyse the reward. The purpose of a reward is to satisfy a craving. Indeed rewards provide benefits of their own. But often what we crave is not the habit itself but the change in mood it delivers, for example smoking brings relief, walking brings calmness, brushing teeth brings fresh breath, and television brings mindless relaxation. Most cravings are obvious in retrospect when they are conquered. At first this can be tricky, and so we might need to observe our own behaviour through reflection and self-questioning. While some people will argue that procrastination is a result of perfectionism – and there may be an element of truth in this – perfectionism may also be a symptom rather that the reward.

Every habit produces a corresponding belief. For example, a chronic procrastinator may say 'I work better under pressure'. However, procrastination is often about not trusting ourselves and being convinced that any effort on our own behalf would be second rate. The fear of doubting our ability to complete a task is one sure way to make certain we put off starting it. Each time we procrastinate, we reinforce a lack of trust in ourselves and affirm feeling helpless and out of control. The good news, though, is that procrastination, like any habit, is learned behaviour. Thus it can be unlearned.

Unlearning Habits

According to a study published in the *European Journal of Social Psychology* (2009), it takes between 18 and 254 days to form a new habit. Within this timeframe, it usually takes 28 days to get rid of an old habit; 26 days for a new habit to become a routine and 99 days for the basal ganglia in the brain to form a new neural pathway for the habit to become automatic. We have the power to take control of the processes that take place in our brain. And so now that we know that the brain is just following our lead, and doing what it thinks is good for us, we can re-programme it. Just willing to change is not enough because it does not address the underlying benefit of the behaviour we want to replace. Evidence for this comes from studies on willpower. However, willpower it seems comes in limited supply. The more we use it, the more it gets depleted, which means we are more likely to give up on later attempts. When willpower is depleted – in times of stress or exhaustion – we fall back on habits. For this reason, unlearning an old habit and replacing it with a new one takes time.

So why do old habits die hard? As we have seen the three-step pattern is the backbone of every habit, and our brain runs through these steps in the same way as we formed a habit as it does to change a habit. To change a healthy habit, the trick is to replace the routine in the loop with something else which will also give you a similar reward. Our brain will think this loop works because we also feel a sense of accomplishment. As we do it more often and experience the same reward in the loop, a new habit will form.

When a habit is in the early stages of formation, the brain's reward response – essentially a dopamine rush (the brain's pleasure neurotransmitter) – is activated once the habit's reward is received. As the habit becomes more engrained this changes. Eventually, the reward response happens right after the cue triggers the habit. It happens in anticipation of the reward. This creates a craving to get that reward and serves to keep the habit going strong.

Some habits once built tend to help people build other habits. To change an existing habit, it is usually helpful to start another habit – to give up smoking, take up jogging. These are called 'keystone habits', and they do this by helping people create small wins in their lives. These small wins increase motivation, foster momentum and create structure for building other habits. When we achieve that small first step, we get a boost of confidence. The keystone habit metaphor is taken from an architectural term referring to the key block in an arch that holds all the other stones in place. Keystone habits are a very effective change technique in the habit-loop strategy.

Educational scholars propose using reflective practice as a means to respond differently in a given situation, or to build a new habit. By engaging in a process of reflection, we accept a continuous cycle of self-observation and self-evaluation in order to understand our actions and the reactions they evoke in us, and in our learners (Brookfield 1995; Mezirow 1990). The reflective process is triggered by the acknowledgement that there is always some aspect of our behaviour that requires attention. Reflection helps to pull us out of 'auto-pilot' and cause us to focus on some part of teaching (Eurat 2002).

One framework which can be useful to guide reflective thinking is Gibbs' reflective cycle (1998). The cycle has six steps: description, feelings, evaluation, analysis, conclusion and action. By working through each of these six stages, Gibbs' framework encourages us to think about the different aspects of a given situation or event, to evaluate it, and establish an action plan for dealing with such a scenario, should it arise again. We are enabled to change our automatic response, and become more conscious of the behaviour we desire or expect from ourselves.

Conclusion

As we have described in this chapter, habits are defined broadly as any learned mental action that can be performed in certain situations without thought. The starting point of changing a habit is to change behaviour by taking action. Changing our brain processes and our old habits requires a shift away from a neural pathway that is undermining one's brain health to a more desirable pathway that works to 'keep our brain busy' and thus optimising its functioning.

Establishing positive learning habits, such as setting aside dedicated time each day for learning, helps create a structured learning environment. The habit of reflective practice for example, can help us to regularly and systematically examine our learning, identify strengths and areas for improvement, and set goals for future development. By cultivating positive learning habits, individuals can enhance their learning outcomes, achieve greater proficiency in their areas of interest thus optimising brain efficiency.

References

Baumeister, R. and Tierney, J. (2012). *Willpower: Why Self-Control Is the Secret of Success.* London: Penguin Books.
Beilock, S. (2015). *How the Body Knows the Mind.* London: Little Brown Book Group.
Brookfield, S. (1987). *Developing Critical Thinkers: Challenging Adults to Explore Alternative Ways of Thinking and Acting.* Milton Keynes: Open University Press.
Corballis, M. (2014). *A Very Short Tour of the Mind: 21 Short Walks around the Human Brain.* London: Duckworth Overlook.

Damasio, A. (2008). *'Descartes' Error: Emotion, Reason and the Human Brain*. New York: Random House.

Duhigg, C. (2012). *The Power of Habit: Why We Do What We Do and How to Change*. London: Random House.

Eagleman, D. (2011). *Incognito: The Secret Life of the Brain*. London: Pantheon.

Eraut, M. (2002). *Developing Professional Knowledge and Competence*. London: Routledge Famer.

Gallwey, T. (1972). *The Inner Game of Tennis: The Ultimate Guide to the Mental Side of Performance*. New York: MacMillian.

Gibbs, G. (1998). *Learning by Doing: A Guide to Teaching and Learning*. London: Further Educational Unit.

Graybiel, A. M. (2008). Overview at Habits, Rituals, and the Evaluative Brain. *Annual Review of Neuroscience*, 31, 359–387.

James, W. (1890). *The Principles of Psychology*. New York: Henry Holt.

Lally, P., van Jaarsveld, C. H. M., Potts, H. W. W. and Wardle, J. (2009). How Habits Are Formed: Modelling Habit Formation in the Real World. *European Journal of Social Psychology*, 40, 998–1009. <https://doi.org/10.1002/ejsp.674>.

Mann, H. (1877). *A Few Thoughts for Young Men*. Boston: Lee and Shephard.

Mezirow, J. (1990). *Fostering Critical Reflection in Adulthood: A Guide to Transformative and Emancipatory Learning*. San Francisco: Jossey-Bass.

Miller, G. A. (1956). The Magical Number Seven, Plus or Minus Two: Some Limits on Our Capacity for Processing Information. *Psychological Review*, 63(2), 81–97. <https://doi.org/10.1037/h0043158>.

Neil, D., Wood, W. and Quinn, J. (2006). Habits – A Repeat Performance. *Current Direction in Psychological Science*, 15(4), 198–202.

Staunton, D. (2012). *Going to College as a Mature Student: The Next Step in Your Academic Journey*. University College Cork: Adult Continuing Education.

Van Blerkom, D. L. (2003). *College Study Skills: Becoming a Strategic Learner* (4th edn). California: Wadsworth.

SECTION 3

CHAPTER 8

Learning across the Life Cycle

> 'Change is a form of new learning, and learning, at the microscopic level is about making new neurons'.
>
> (Banks 2015: 79)

In this chapter we will discuss how the brain is built, from the moment of conception to old age, through an ever-changing cycle of adaptation and change. When we deal with a system as complicated as the human brain, we cannot expect simple answers. Nor can we hope to comprehend such sophisticated mental processes as attention, awareness and consciousness simply by seeing which parts of the brain 'light up' during scans, especially as a person performs all manner of tasks.

One of the main characteristics of the brain is the ability to learn and to form memories. For most people, their memories define who they are, where they have come from and who they want to be. Learning and memory are two ways of thinking about the same aspect of the brain: both describe an outcome based on the ability of the brain to change its structure and its function because of experience. After all, memory is learning that persists.

Educationalists agree that learning is a process that occurs over time; it is not instantaneous or a straightforward progression in a straight line. Most learning usually involves at least two different types of memory. Implicit learning involves the actions, procedures and skills that we have internalised. They have become ingrained, unconscious habits. For the first eighteen months of our lives, we only encode implicit memory: smells, taste, sounds, bodily and emotional sensations. This information is embedded in our synaptic connections and becomes unconscious memories which prime the brain to be able to do things.

When we have mastered tying our shoelaces for example we will likely be able to secure our shoes properly for the rest of our lives. This durability is true of all learned behaviour in virtually every aspect of life, from riding a bicycle, reading a book or people's facial expressions, to participating in a conversation. Implicit in this type of learning is the notion that relatively long-term change in how we act and think results from experience.

In contrast explicit memory involves the type of learning we experience in school when we learn about such thing as facts, events, people and cultural expectations. Normally we need to make a conscious effort to recall these memories (such as our first kiss). Neuroscientists have made great strides in uncovering some of the secrets in a human brain and although it is constantly adapting and growing throughout the lifespan. In this chapter we will focus on the changes that occur during infancy, adolescence, mid-life and old age.

Building a Brain during Infancy

Until very recently, it was believed that in humans, neurogenesis happened only during gestation and that by the time babies were born they had their full quota of neurons. However, in 1998, a landmark piece of research showed that the adult human brain was capable of producing new neurons in the hippocampus (Kempermann and Gage 1997), the region of the brain responsible for learning. Subsequent studies have found that we are also able to grow neurons in the cerebellum throughout the life cycle. There are four processes involved in the phases of building a brain: neurogenesis, synaptogenesis, myelination and pruning.

Neurogenesis refers to the birth of new neurons. As we age, the structure becomes so complex and finely tuned that replacing lost neurons with new ones would disrupt the system. Thus there is good reason for a general lack of neurogenesis in the human brain following infancy. The bulk of the brain-building process occurs through synaptogenesis, which is the development of new synaptic connections and myelination. Myelin sheaths allow axons to remain thinner but maintain a faster conduction speed for the

Learning across the Life Cycle

action potential, when neurons are stimulated to fire, thus providing faster information processing. Although the human brain is constantly adapting and growing throughout life, the biggest bursts in synaptogenesis, myelination and pruning primarily occur during infancy, adolescence and old age. While the brain has reached its full size by the age of ten, it undergoes dramatic organisational change that continues well into early adulthood.

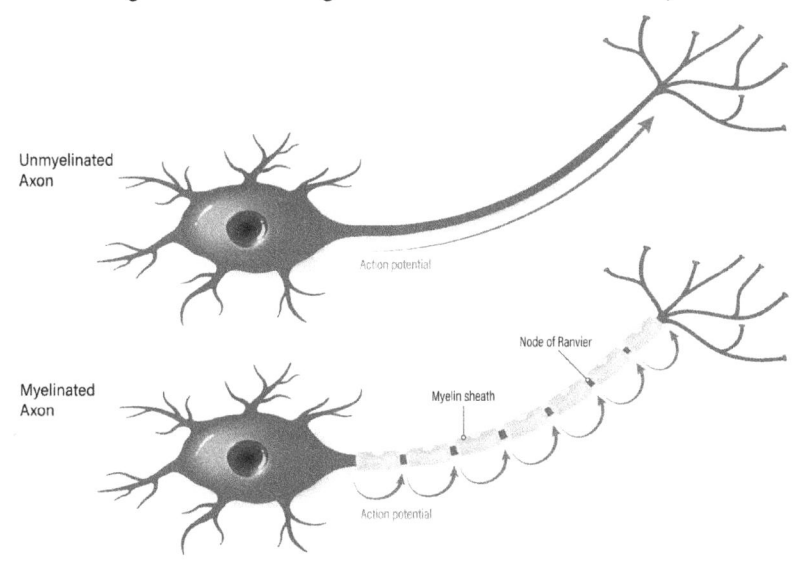

Figure 14: Myelinated and Demyelinated Axons – Image by Dee-sign/<stock.adobe.com>

Brain development was long thought to be complete within the early years of life, but new research shows that the brain continues to mature throughout adolescence and into early adulthood (Merriam and Baumgartner 2020). This results in a prolonged period of plasticity that makes teenagers highly vulnerable to social, educational and emotional challenges during this transition phase. Myelination progresses like a wave that gradually sweeps through the brain, starting at the back of the organ to the front. Thus, areas near the back of the brain, such as those involved in vision, are upgraded relatively early on, while the prefrontal cortex, located behind the eyes does not reach full maturity until the late 20s or 30s.

The last but equally important part of the brain-building process is pruning. This involves getting rid of neurons and synapses that are not needed. Although it may seem counterintuitive to kill off perfectly healthy neurons, the mechanism is just the same as a rose bush or tree will grow better and stronger flowers after pruning,

Brain Development in Childhood

No other organ in the human body takes as long to develop or goes through as much change as the brain. A child's brain begins to form from the second or third week of foetal development and continues well into early adulthood. In the 9 months from conception to birth, the brain develops at a phenomenal rate, during its peak a quarter of a million new brain cells are born every minute (Bloom et al. 2003).

Within 24 hours of fertilisation, cell division has begun. In the third week after fertilisation, the neural tube forms. The neural tube is the beginning of the central nervous system. The front end of the tube develops into the brain and the remainder becomes the spinal cord. At around this stage, the cells of the neural tube begin to divide very rapidly. Some of these dividing cells become neurons (the basic cellular unit of the brain) and some will become glia (a range of cells that support the emerging neural network by providing structure and nourishment).

The midpoint of pregnancy is significant as the brain changes from a primitive structure into a more complex form. According to Bloom et al. (2003), cells begin to travel (migration), form communities (aggregation) and make connections that enable the communication necessary to brain function (synaptic formation). The peak growth of neurons and cells occurs in the fourth and fifth months of pregnancy when the cortex grows more rapidly than the supporting structures underneath. By the sixth or seventh month of pregnancy, 70 per cent of the brain's neurons are located in the cerebral cortex.

During the final stage of pregnancy, two new processes begin: pruning of unnecessary cells and the protection of vulnerable cells known as myelination. Over the next few weeks and months, as the cells continue to divide and move, the wall of the neural tube thickens and forms the structures of the cerebral hemispheres, cerebellum, brain stem and spinal cord. Once the migrating neurons reach their destination, they then begin to send out dendrites and axons.

At around the fourth month of pregnancy, synapses begin to form. This is when the brain asserts its uniqueness. A single neuron may have thousands of synaptic connections, meaning that within the developing brain, there are many trillions of connections between neurons. As the brain continues its development, it produces more neurons and glial cells than it will need.

At birth a child's brain will consist of 100 billion neurons, grouped into sets that execute specific functions such as interpreting sounds, storing memories and learning new skills. Then in a process known as pruning, half the cells die off. During infancy the connections among neurons become stronger and more widespread, with new connections forming. These connections form the networks that underlie children's development. Some neurons have a function that is temporary and to do with development itself, thus they are no longer needed when their job is complete. Other neurons are created for insurance purposes, whereby they stay only as long to guarantee that all the proper connections are in place. When the proper connections are all ready for that development stage, any extra neurons are pruned away.

The First Five Years

In the first five years the brain reflects the demands placed upon it. Its development is based on both biological growth and social stimulation and is primarily driven by sensory, emotional and social experience. This means that certain things are programmed to happen while other things depend on external stimulation. As soon as they are born, babies begin scanning the world around them, looking for meaningful patterns. In this way, infants are learning perceptual development, as the perceptual system recreates the surrounding environment in the brain, based on information provided from the senses (Berk 2000). Both nature and nurture play a significant part in all these developments and neuroscientists often compare the building of a human brain to the weaving of cloth: some threads are supplied by genes, with others provided by the environment. In the resulting fabric, the different strands are so tightly woven they are virtually indistinguishable.

It is said that in the first few years of life, 700 new neural connections are formed every second. Neuroscience research has revealed that the 'synapses in a child's brain multiply twenty-fold between birth and 3 years of age, a rate that is faster than in any other time in life' (Gerhardt 2005). At 3 months, a baby is still quite foetal, yet its brain will grow about 20 per cent in the following 4 to 5 months. Experiences of the world are what keeps a baby growing and developing properly. When a baby is born, each cell of the brain has around 2,500 synapses, which are the connections that allow the brain to pass along signals. In the next 3 years, that number grows to about 15,000 per brain cell, as the brain creates 700–1,000 new neural connections every second. Synapse formation for key developmental functions such as hearing, language and cognition peak during this time, making this window of opportunity in a young child's life extremely crucial for the development of higher-level thinking in later years (Piaget 1964).

During the first 5 years, the stages of peak development are windows of opportunity. These are crucial stages for the brain to be exposed, through the sensory systems, to sound, to sight, to shape and colour, to movement, to tastes and to smells. Missing these windows may cause problems later in life (Bloom 2003). These ideal sensitive periods remain open for a limited time and the first 5 years of life contain most, but not all, of these periods. For example, visual acuity is acquired in the first 2 years of life. This window of time is crucial in the creation of properly functioning eyesight, which is a factor in facial recognition and attachment, and binocular vision which is a factor in acquiring reading skills (Bowlby 1969).

The most considerable growth in the brain connections associated with language development peaks before 12 months. Early brain research has revealed that by the age of 1, the size of the child's brain is already 72 per cent of adult volume and by age 2 it has grown to 83 per cent, with a child's vocabulary expanding from 55 words at 16 months, to 225 words at 23 months to 573 words at 30 months (Goswami 2015b). This growth, to some degree, is dependent on the experiences the baby receives. While some level of language development occurs naturally, fuller development of children's language capacities requires the engagement of knowledgeable conversant adults and the encouragement of children's verbal expression. For Vygotsky the adult role is critical to the quality of play and learning for the child while according to Aikin, 'effective oral language skills are the building blocks on which subsequent literacy and numeracy is based. Without solid foundations in language and communication skills, children run the risk of school failure, low self-esteem, and poor social skills' (2014: 44).

Adolescent Brains Are a Work in Progress

Adolescence is normally viewed as a time of transition from childhood to adulthood and characterised by rapid mood swings, drama, sulks, experimentation and acute self-consciousness. Starting at about age 11 or 12, the brain undergoes major reorganisation in areas associated with managing impulsivity, social interaction and risk assessment.

The transition from childhood to adulthood involves an interaction between two neurological and psychological systems. The first concerns the limbic system, which is involved in emotion and motivation. Research shows that the nucleus accumbens or 'reward centre' in the brain is more active in adolescence than in children or adults. This has important implications for young people and goes some way towards explaining some of their risk-taking behaviour. Contrary to popular belief, teenagers overestimate the risks associated with destructive conduct, but because their brains are particularly sensitive to dopamine, they also overestimate how rewarding such behaviour will be.

The other system is the prefrontal cortex, the brain's control system, which is involved in complex functions such as decision-making, long-term-planning, self-regulation and delayed gratification. Brain imaging technology shows that throughout adolescence the brain upgrades its wiring by increasing its production of myelination. At the same time, connections throughout the cerebral cortex are continuously being refined by experience. There is an ongoing interaction throughout the lifespan between the limbic system and prefrontal cortex but during the adolescent stage the limbic system has a greater influence. However during adulthood the prefrontal cortex becomes more dominant.

Teenagers learn from their mistakes, through trial and error, and this influences the development of the prefrontal cortex so that their control system improves. This is facilitated by the ongoing processes of synaptic formation and myelination, which lead to faster connections and increases the brain's networking to improve the control system. And the creation of stronger connections between the prefrontal cortex and the hippocampus means that learning from past experiences is increasingly integrated into the decision-making process.

It has been assumed for many years that, by the time a child begins secondary school, all the hard-wiring in the brain is complete. Yet for example according to research by Smith, adolescents appear rather poor at reading the emotions on the faces of individuals whose photographs they were shown. In many instances completely wrong interpretations were made - as anger was confused with joy, sadness with fear. Their brains were not yet wired by experience of a range of social interactions (2002: 51). More recent research findings show that the time around puberty is critical for the trimming of neural connections, especially in the pre-frontal cortex as it begins to take more control (Spinks n.d.). Brain growth continues to the age of twenty-five whereupon it reaches maturity (Johnson and Siegel 2010).

> **Case Study**
>
> *BrainPower* is an open access course that was developed by a team of academics and scientists from occupational science and occupational therapy, applied psychology and neuroscience, with contributions from students and higher education staff, in University College Cork and funded by the National Forum for the Enhancement of Teaching and Learning in Ireland. It is a free, self-paced course, focusing on harnessing the power and potential of adolescent brain and behaviour for enhanced learning, wellbeing and student success in higher education.
>
> The course consists of six modules, each of which describes brain development as it relates to students in higher education settings, identifying risks and opportunities of socio-cultural contexts relating to student learning in higher education settings, and exploring practical strategies for enhanced student success and the development of graduate attributes through curriculum design, learning and teaching. Taking an interdisciplinary approach, it tackles brain health and wellbeing from various perspectives and is accessible to a general audience. Although targeted mainly at academic and academic support staff in higher education, Brainpower is likely to be of interest to staff across further and higher education, as well as parents, school teachers, coaches and all those who contribute to the lives of young people.
>
> Available from: <https://www.ucc.ie/en/cirtl/professional-development/badge/brainpower/>

The Adult Brain

If the hallmark of the child's brain is 'wiring up', and the greatest feature of the adolescent brain is frontal lobe development, the most important trait the brain brings to adulthood and through the end of life can be summed up in one word: plasticity. As we have seen plasticity is the term neuroscientists have coined to describe the brain's biological adaptation in response to experiences and change. It allows us to learn, to unlearn, to form new habits and to adjust to new circumstances such as negotiating new relationships, rearing children or learning to use information technology.

The adolescent years are a time of challenge and growth, when the young person is forming their identity. Middle-age is characterised as a time of re-evaluation and self-reappraisal (Livingstone 1999). As adults we have the personal freedom to make conscious choices and take purposeful steps to steer ourselves in a desired direction. Adults aged 40 through to 65 years, may go through a 'second adolescence' struggling with their own ambivalence about getting old (Neugarten 1976; Wright 2017). In the mid-years, adults re-evaluate their identity, asking: What have I done with my life? What is it I want for myself and others? What are my central values? Most people experience this period as a quest, a desire to answer deep questions and search for what is true and meaningful in their lives.

The grappling with new dilemmas during the middle years in the life cycle provides the adult brain with the push for developmental intelligence – the maturing of our individual capacities for cognition, emotional intelligence, judgement, social skills, life experience and consciousness (including spirituality). Developmental intelligence is expressed in deepening wisdom, judgement, perspective and vision and is evident in three different types of thinking (Perry 1970), including:

1. Relativistic thinking: knowledge is not neutral, can be biased in favour of a particular ideology or context. It is not absolute, nor does it rest with authority (people or institutions).
2. Dialectic thinking: the ability to uncover and resolve contradictions in opposing and seemingly incompatible views. Dialectical thinking is 'finding our own voice' based on experience, reflection and insight.
3. Systematic thinking: being able to see the larger picture by distinguishing between the important and the unimportant through logical reasoning.

These new ways of thinking, in turn, trigger new dilemmas during middle-age (Mezirow 1990; Brizendine 2010; Wright 2017):

1. Perspective on time changes: up until mid-life life seems like an endless upward slope, with nothing but the distant horizon in view. Now there is the experience of reaching the top of the hill, with the end of the road in sight, but the horizon still far away.

This marks a turning point with a sense of impatience to do new things and achieve unfulfilled aspirations or desires.
2. A balance between being young and old becomes more apparent and this awareness brings to the surface a new set of questions about the notion of liberation and freedom. We give up aspects of being young, naïve and insecure and we take on aspects of being reliable, competent self-aware adults. We do not view middle-age as stagnation but wish to become productive members of society and give something back to the next generation.
3. A new balance between destructiveness and creativity: by mid-life, everyone has experienced human destructiveness. In the face of the hurt caused by self and others, we have to give up the myth that life is simple and that there are dark forces in the personality. We have to come to terms with grief over lost opportunities, rage over betrayal of self. Mid-life calls for a new relationship with this dark side to create or bring something new into being.
4. A balance between need for attachment and separation: in a paradoxical way, as the middle-aged person takes on more responsibilities, there is a corresponding need to turn inwards and engage with the self. The result is that a person can draw more on inner resources.
5. A new balance between the masculine and the feminine aspects of personality: for those who have concentrated on the more masculine aspects of their gender identity, there is the need to attend to the more feminine aspects and vice versa. When they are combined, people can be more creative in work and relationships (Gilligan 1982).

Expert Insight

Dr Harris Eyre, Fellow in Brain Health at the Centre for Health and Biosciences

In conversation with the authors, Dr Eyre, who is co-founder of the PRODEO Institute, and advisor to the Organisation for Economic Co-operation and Development (OECD), highlighted some of the key requirements to move educational neuroscience forward to support learning across the lifecycle.

Transdisciplinarity is essential for educational neuroscience

To really progress the links between neuroscience and education, we need transdisciplinary thinking. That is taking every discipline, and intentionally forcing them together. You know that people are going to be frustrated, they're not going to understand each other. But they all need to have the right mindset because we need to develop a new vernacular, a shared understanding. It's going to create fireworks and friction, but it's going to come to something good. The brains and minds of society are really struggling. We have rising rates of mental health and neurological issues. We don't have an adequate number of clinicians to help people. We need technological solutions but they don't exist. We need to develop a completely new industrial innovation strategy around brain tech, brain tech for mental health, brain tech for neurology, brain tech for Ed Tech.

A key neuroscience innovation that supports education is brain capital. Brain capital is a construct. It's not a technology or a foundational engineering tool. It's an economic construct that basically, the argument is that all of the stresses to people's brains that are occurring in the modern economy, you know, financial stress, climate stress, disinformation, stress, misinformation, stress, all of these things are impacting our brain. So, brain capital, which incorporates brain health, across the lifespan and brain skills across the lifespan, is an organising construct and an economic asset that I would argue to say that we need to govern our nations based on brain capital as a key pillar, you know, at the same level of significance as GDP, and road and bridge infrastructure and clean water.

The school is the bedrock for identity formation and mental immunity

We need to recognise that people in the education system are vital to nurturing and building the brain capital of the future society. Children spend more of their time in school, and it's so complicated with neurodiversity, mental health issues, geopolitical issues, climate change etc. but the school is like the bedrock foundation of forming children's identity and helping them prepare themselves for success. One of the biggest neuroscience advances is the idea of mental immunity. It is particularly important for teenage and adult learners when accessing information digitally and through social media. If we use the metaphor of disinformation as a virus, it scrambles your mind or your sense of reality. So how do we build up antibodies and our immune system of our brain to sort of quickly recognise disinformation, and to disregard it so it doesn't get and invade our brain? Learners need to be taught how to engage critically and to ensure that their brains are switched on to disinformation. Children also need to be exposed to neuroscience and psychology and brain health, to be ready to engage critically with AI. New cartoons are being created like *Brain Robbie* which talk to schools about brain health and food, social engagement and physical safety.

Recent Research

Ibanez, A. and Eyre, H. (2023). Brain Capital, Ecological Development and Sustainable Environments. *BMJ Mental Health* 26, 1–3.

Eyre, H., Lundin, R., Falcão, V. Berk, M., Hawrot, T., Leboyer, M., Destrebecq, F., Sarnyai, Z., Reynolds, C., Lavretsky, H., Kolappa, K. and Cummings, J. (2023). Brain Health Is a Determinant of Mental Health. *The American Journal of Geriatric Psychiatry*, 31.

Eyre, H. A., Ayadi, R., Ellsworth, W., Aragam, G., Smith, E., Dawson, W. D., Ibanez, A., Altimus, C., Berk, M., Manji, H. K., Storch, E. A., Leboyer, M., Kawaguchi, N., Freeman, M., Brannelly, P., Manes, F., Chapman, S. B., Cummings, J., Graham, C., Miller, B. F., Sarnyai, Z., Meyer, R. and Hynes, W. (2021). Building Brain Capital. *Neuron*, 109(9), 1430–1432.

The Older Brain

Just as ageing brings risks and predictable changes to the brain, it also opens new possibilities and consolidates the gains of a lifetime, a reason for gratitude. A study that followed more than 200 people over a 50-year period found that the majority saw their psychological health improve – not simply happiness but the qualities of being dependable, productive, responsible, and of having good relationships with people increase steadily with age (Buettner 2008). This happy flourishing suggests ongoing processes in the brain which may help us understand how the accumulation of experience and the passage of time can work together to produce one of the most prized qualities of all – wisdom.

In stark contrast to this dementia is undoubtedly the most feared outcome of old age. In the last decade the concept of dementia has undergone a dramatic change, because there is simply no general condition that can be called dementia. Previous generations, unable to recognise the very gradual development of brain degenerative diseases, had to assume that what they were observing was the typical course of old age. Neurodegenerative diseases, if they are to occur at all, do tend to appear at a more advanced age. However seen in terms of lifelong development, the subtle attrition of neurons with age may represent not a sad loss but a progressive fine-tuning of cerebral networks.

We learn best in the stretch zone, when we have some control over everyday events and find a sense of meaning that extends beyond our day-to-day life. Purposeful learning for brain health entails pushing ourselves out of our comfort zone and continuing to challenge ourselves as we age. To be able to throw ourselves into a cause for example and strive to attain a meaningful goal is what psychologists call the 'flow'. In this moment of intense presence there is no sense of past or future, just an overwhelming feeling of being 'in the moment'. It seems that engagement in what we are doing, whether it is work or leisure, seems to be especially important for happiness and the promotion of cognitive reserve.

Lifelong Learning and Cognitive Reserve

Scientists have long been intrigued by a group of people who show little if any mental deterioration until just before they die. They appear to be protected from the mental decline that comes with age. Autopsies on such people often reveal their brains to be riddled with the plaques and tangles characteristic of advanced Alzheimer's disease. We mentioned one such group earlier when talking about the Nun study. However many elderly people today continue to retain their full mental capacities compared to other people of the same generation. To explain this striking difference, researchers came up with the 'cognitive reserve' hypothesis, according to which our brains have some degree of reserve capacity.

It appears our brains can develop some degree of reserve capacity through lifestyle choices. The theory suggests that with increased challenge and new learning, the brain reacts by literally developing more synaptic density (greater number of connections between neurons) and greater volume, thereby offering a better defence against loss of mental acuity. Therefore learning appears to be an important health promoting behaviour that contributes to brain reserves. It is important to note that early years of school in life are the typical measure used to identify level of education and qualifications achieved. However lifelong learning and lifespan mental stimulation are also considered important contributors to brain reserve.

In the book *Ageing with Grace*, Snowdon (2002) relates the story of a conversation with Sister Nicolette (then aged 91), in which he asked her to what she attributed her good health and longevity. She replied that she continued to stretch her brain through reading theology and had an exercise programme which involved walking several miles a day. When asked when she started this programme, Sister Nicolette replied 'When I was 70'. Cognitive function is modifiable throughout life, and it is never too late to take control. The best advice is to learn something new and novel. Staying mentally active helps to cushion people who have started to suffer age related decline. If we do not use it, we lose it, as the popular phrase goes.

Conclusion

In this chapter we have demonstrated that brain research consistently proves the law of exertion: if we want something to get stronger, we must exercise it. If we want to keep our mental acuity, then we must use our mental abilities regularly. Throughout the life cycle, learning is not instantaneous, nor is it a straightforward progression in a straight line. Learning is influenced by individual differences, cultural factors, socioeconomic status, and access to educational resources, all of which impact brain development.

One of the great insights from neuroscience research points to the fact that the older brain is as capable of change as the younger brain. As emphasised in this chapter, cognitive reserve theory demonstrates that learning something novel and new in old age has a powerful impact on preventing brain degenerative diseases. Furthermore, industries are constantly evolving with advancements in technology, changes in regulations, and shifts in best practices. In adulthood building on our strengths and engaging in continuous professional development can not only ensure personal growth, but benefit career development in a rapidly changing work environment.

References

Aikin, M. (2014). *The Cyber Effect: How Human Behaviour Changes Online*. London: John Murray Publishers.
Banks, A. (2015). *Wired to Connect: The Surprising Link between Brain Science and Strong, Healthy Relationships*. New York: Penguin Random House.
Berk, L. (2000). *Child Development* (5th edn). Needham Heights, MA: Pearson.
Bloom, F., Beal, F. and Kupfer, D. (Eds) (2003). *The Dana Guide to Brain Health: A Practical Family Reference from Medical Experts*. New York: The Free Press.
Bowlby, J. (1969). *Attachment and Loss*. New York: Basic Books.
Buettner, P. (2008). *The Blue Zones: Lessons for Living Longer from the People Who Have Lived the Longest*. New York: National Geographic.
Giddens, A. (1991). *Modernity and Self-identity: Self and Society in the Late Modern Age*. California: Stanford University Press.
Gilligan, C. (1982). *In a Different Voice: Psychological Theory and Women's Development*. Boston: Harvard University Press.
Goswami, U. (2015a). Sensory Theories of Developmental Dyslexia: Three Challenges for Research. *National Review Neuroscience*, 16(1), 43–54. <https://doi.org/10.1038/nrn3836>.
Goswami, U. (2015b). *Children's Cognitive Development and Learning*. Cambridge: Cambridge Primary Review Trust.
Johnson, S. and Siegel, H. (2010). *Teaching Thinking Skills*. London: Continuum International Publishing Group.

Kempermann, G. and Gage, G. (1997). More Hippocampal Neurons in Adult Mice Living in an Enriched Environment. *Nature*, 386(3), 493–495.

Livingstone, D. W. (1999). Exploring the Icebergs of Adult Learning: Findings of the First Canadian Survey of Informal Learning Practices. *Canadian Journal for the Study of Adult Education*, 13 (2), 49–72.

Merriam, S. and Baumgartner, L. (2020). *Learning in Adulthood: A Comprehensive Guide* (4th edn). San Francisco: Jossey-Bass.

Mezirow, J. and Associates. (1990). *Fostering Critical Reflection in Adulthood: A Guide to Transformative and Emancipatory Learning*. San Francisco, CA: Jossey-Bass.

Neugarten, B. (1976). Adaptation and the Life Cycle. *Counseling Psychologist*, 6(1), 16–20.

Perry, W. (1970). *Forms of Intellectual and Ethical Development in the College Years*. Austin, TX: Holt, Rinehart and Winston.

Piaget, J. (1964). 'Part 1: Cognitive Development in Children: Piaget Development and Learning. *Journal of Research in Science Teaching*, 2(3), 176–186.

Smith, A. (2002). *The Brain's Behind It: New Knowledge about the Brain and Learning*. Stafford, UK: Network Educational Press.

Snowdon, D. (2002). *Aging with Grace: The Nun Story: Living Longer, Healthier, and More Meaningful Lives*. London: Bantam Books.

Sousa, D. A. (2001). *How the Brain Learns* (2nd edn). London: Sage Publication.

Spinks, S. (n.d.) Adolescent Brains Are Works in Progress. *Frontline Inside the Teenage Brain*. Available from: <https://www.pbs.org/wgbh/pages/frontline/shows/teenbrain/work/adolescent.html>

Wolpert, L. (2011). *You're Looking Very Well: The Surprising Nature of Getting Old*. London: Faber and Faber.

Wright, R. (2017). *Why Buddhism Is True: The Science and Philosophy of Meditation and Enlightenment*. New York, NY: Simon and Schuster.

CHAPTER 9

Interpersonal Relationships

'Through others we become ourselves'.

(Vygotsky, 1978 cited in Flanagan, 2021: 85)

We are deeply social creatures. How inter-personal relationships benefit the brain are now better understood. Humans are in large part driven by the need to belong, to seek out and cultivate meaningful emotional relationships with others. For our brains to thrive, we need interpersonal connections. Staying socially active and connected regularly with family, friends, colleagues and other groups, the better we may become at preserving mental sharpness. Education can provide a safe environment in which these relationships are formed, where young people and adults can share stories, confide in each other and open up to a world of new experiences.

Interaction with others refers to our interpersonal relationships, our social activities, the extent of our social networks, and our functional and emotional social supports. Historically the brain has been studied in isolation, but that approach overlooked the fact that an enormous amount of brain activity has to do with our involvement with others. Our social interactions play a crucial role in developing our sense of self, our memories, our relationships and friendships, as well as the outcomes we create. If managed well, social interactions enhance our well-being and the quality of our overall health. If anything, recent experience with COVID-19 has highlighted the crucial role of human connection in our sense of wellness.

In this chapter we will examine why humans are driven to cultivate meaningful emotional relationships with others. In this context we will discuss how strong teacher relationships can be a protective factor for students' social and academic development. In other words, why do we have a social brain? These interactions stimulate the brain and require a complex combination of skills including attention, memory, thinking, speech and social awareness. There is now a surprising link highlighted by science between strong healthy relationships and our brain health. A rich social network provides sources of support, reduces stress, combats depression and enhances intellectual stimulation. Conversely, if our social interactions are managed badly or inappropriately, they can have a negative impact on both how we feel and how we perform. Socialising is probably one of the most enjoyable ways to maintain brain health and cannot be underestimated. The benefits of social interaction are compelling, with those with strong relationships appearing to manifest less cognitive decline and living more active, adventurous lives.

The Social Brain

What does our brain need to function normally? Beyond the nutrients from the food we eat, beyond the oxygen we breathe, beyond the water we drink, there is something else, something more critical to survival: social life. Staying connected with others makes us feel more connected to the world and increases our sense of belonging. Normal brain health function depends on the social web around us. We may not absolutely need one another to survive but we certainly need other people to thrive.

Early on in our history as a species, we not only survived, but prospered by gathering together. The 'social brain hypothesis', proposed by anthropologist Robin Dunbar, argues that our ancestors developed cooperative social skills which allowed them to survive and thrive in many diverse physical environments and evolve ever-larger social groups. It was the gathering together into social groups that provided mutual protection and assistance. This early experience resulted in the evolution of the complex sprawling networks in the brain that enable us to recognise others, monitor their intentions and guess their desires and beliefs, as well as feel their pain or pleasure and read their emotions and actions. We are, as brain scientists continue to remind us, wired to connect.

The complexity of life in groups is strongly determined by social skills deeply rooted in our neural circuitry. Neuroscientists inform us that the animal brain becomes highly adaptable to the particular habitat in which it lives through learned behaviour. These behaviours become imprinted on the brain during the early stages of life and are reflected in strongly instinctive ways of coping. We humans are distinguished by the fact that we come into the world with a particularly 'unfinished' brain that remains susceptible to formation by experience, and by the fact that we live in groups with a shared culture (Cozolino 2014). Consequently, a great deal of time is devoted to the socialisation of children and adolescents in learning social communication skills through education (formal and informal).

A bonded society is made up of layers, like an onion, with one's best friends at the core and successive layers filled with individuals with whom we are decreasingly intimate. Our relationships with other members in our social group can be ranked on a simple continuum based on frequency of interaction and emotional closeness. Since the quality of a relationship depends on the time invested in it, and there are only so many hours in a day, this sets an upper limit on the number of friends we can have. According to Dunbar, the core tends to consist of about 5 intimates, with the next layer takes the group to about 15, while the widest circle encompasses about 50 friends and acquaintances. Based on the social brain hypothesis, it has been calculated that our social group size is around 148 (Dunbar 2011). This turns out to be both a common community size in human social organisations and the typical size of personal social networks, even on social media.

The idea of the 'social brain' suggests that humans must learn to maintain two distinct kinds of relationships simultaneously: intimate relationships with principal partners and friends, and weaker ones with other group members. Relationships are fluid and they follow a model of concentric circles that can be drawn into three categories: relationship with ourselves, with others, and with the wider world. We increase our focus on each of the circles, depending on what stage of our lives we are in.

Being socially engaged triggers physical and emotional changes in the brain. Conversations have the power to change the brain by boosting the production of hormones and neurotransmitters. Social interaction stimulates both physical systems and neuronal pathways thus changing the body's chemistry and the connection of new neuron assemblies in the brain. The rhythms of brainwaves between two people taking part in a conversation begin to match each other. This of course depends on the nature of the conversation – observe how young lovers in a restaurant appear to only have eyes for each other. As we communicate higher levels of dopamine and oxytocin are released which in turn, makes us feel relaxed and trusting. This may be a key factor in explaining why special hormones are triggered in the brain.

In addition to conversation and interaction, some of the therapeutic effect of social connection comes from actual physical affection. When we touch someone we love – partner, child, a friend, even a pet – the hormone oxytocin is released into the bloodstream and has a number of effects. It increases relaxation and reduces our stress response. It also increases the desire for more touch and affection and also causes the release of endorphins, which reduce the pain of everything from headaches to arthritis to even migraines. Oxytocin stimulates the release of all the other sex hormones as well (oestrogen, progesterone and testosterone) making a tonic for hormone balance and overall well-being. We can even get an oxytocin surge from petting our cat, dog or horse.

Like dopamine, oxytocin is produced in the hypothalamus, the brain site that maintains the body in a steady state by regulating such features as temperature, blood pressure, thirst and hunger. The hypothalamus functions more or less as a clock and is intimately connected with organs throughout the body and tightly linked with other control systems in the brain. Oxytocin is associated with promoting social affection and most caring behaviours such as parental love, bonding and attachment with others (Nussbaum 2003: 125). In particular, oxytocin appears to promote and propel friendships and the need to reach out to another person, increasing relaxation and the desire for touch and physical affection.

Those who enjoy a lifestyle in which they often interact with others benefit substantially. The healthiest people seem to have most diverse social networks. Interaction with an informal network of people reduces the harmful effects of health risks including cardiovascular problems, high blood pressure and may even prevent infectious diseases (Gawande 2015).

Mirror Neurons and Imitation

In the late 1990s scientists first discovered what they called mirror neurons in brains of macaque monkeys (Gallese et al. 1996). A few years later, other scientists were able to provide direct evidence of mirror neurons in the human brain. Mirror neurons are a particular kind of neuron that 'reflects' the actions of others and they have been linked to many behaviours and abilities. Mirror neurons become active when a person performs an action or when they see someone else performing that same action. This particular kind of neuron is threaded throughout various parts of the brain, particularly in the motor cortex, interpreting and predicting intention from the movements of others.

As part of a research experiment volunteers were hooked up to some electrodes, designed to measure their facial expression. They were asked to watch subliminal flashes of facial expressions on a computer screen. Interestingly even when the images flashed so quickly that they were below the threshold of conscious awareness, the volunteers' facial muscles matched the expressions. When volunteers were asked to rate how they felt, the ones who were subliminally exposed to smiling faces said they felt good, and the ones who were shown the frowning faces felt worse (Lawton and Webb 2017). This experience is referred to as the 'similarity in experience' sensation which triggers a brain pattern similar to our own imagined emotional reaction. It is a spontaneous process which allows us to more accurately feel the sensations and feelings we get in response to others' emotions.

Our mirror neuron system gets more active the more expert we are at an observed skill. When pianists listen to someone else's piano performance, the finger areas in their primary and premotor cortex increase above their baseline activity. Their mirror neuron systems automatically run the performer's key strokes in emulation. The same thing does not happen in the brains of non-musicians. While they can certainly appreciate the music deeply, their experience is inevitably shallower than the pianist's in at least one way, because they are not experiencing what it is like to actually produce it. The same goes for athletes (any sport), the better our own skills, the more deeply we understand the skilled performances we witness

(Macknik and Martinez-Conde 2011). The actions we mirror most vividly are the ones we know best.

Enriched Social Environment

We may not absolutely need one another to survive but we certainly need other people to thrive. Staying connected with others makes us feel more connected to the world and increases our sense of belonging. Social engagement challenges us to communicate frequently and to participate in complex social interactions (listening, story-telling and showing empathy). It contributes to psychological empowerment gained from a sense of belonging and self-esteem (the feeling that we matter to others). Finally, social networks fostered by social engagement can assist in many practical ways including emotional support (giving and receiving love).

As Covey suggests building relationships involves understanding the perspectives, needs, and concerns of others before expressing one's own thoughts – 'seek First to Understand, Then to Be Understood'. Covey promoted the idea of taking personal responsibility for one's actions and contributions to relationships, and building trust through consistency, integrity and keeping commitments. Socialising is probably one of the most fun ways to maintain brain health. Friends can open up our world to new experiences, share the burden of a difficult situation, and make us feel on top of the world. When we feel connected to other people, we have an aim in life and a sense of responsibility. Social interaction boosts brain 'muscles' because it involves many behaviours that require memory and attention.

Research suggests that there is a strong positive relationship between an enriched social environment and social engagement with a wide network of people and contributes to brain functioning and a personal sense of well-being (Kenyon 2015). A rich social environment can be defined as having 6 or 7 social ties with significant other people including: (a) extended family networks, (b) single sex or mixed friendship groups, (c) membership in a leisure/sporting/cultural clubs, (d) religious or spiritual community quest, (e) frequent travel with others, (f) participation in community/voluntary organisations and (g) a meaningful text or e-mail exchange, in additional to an actual conversation at least once a month.

The size of the human brain has tripled in the past three million years and much of this growth is devoted to interpersonal relationships. Many evolutionary processes have shaped our nervous system, giving us the remarkable capabilities to foster bonding with others through cooperative relationships. Over the 100,000 generations since tools were first invented, those genes that fostered relationship abilities pushed their way forward in the human gene pool. We see the results today in the way our human nature is moulded to connect – including caring for children, concern for others, altruism, fairness, generosity, language, forgiveness, morality and religion. While our primary motivation as a species is to survive through food and sex, interpersonal relationships remain a significant factor in the process.

The human parent-child relationship is unique. It has a singular power to shape how each of us pursues and expresses empathy as an adult. No other animal nurtures their offspring for so long as human caregivers do for their children. The helpless child is dependent on adult support for safety and stability, as well as a warm and responsive social and emotional connection. We call this learning about how to bond (Bowlby 1969).

When children experience secure attachments they learn about empathy, what it means, and how to express it from the way they are loved or not loved, responded to or ignored. As a bond between parents and a child becomes deeper, and grows and develops, the ability to feel attachment will gradually be generalised from parents to extended family, to wider networks and friends. Thus healthy secure attachment begets caring and caring begets empathy and empathy begets compassionate action.

The neural pathways laid down in the early months and years of life provide the architecture for the future cognitive and emotional development of children. A major factor in children's brain development is what has been termed the serve and return interaction between children and their caregivers (Shonkoff 2017). Young children naturally reach out for interaction through babbling, facial expressions and gestures, and adults respond with the same kind of vocalising and gesturing back at them. This back-and-forth process is fundamental to the wiring of the brain, especially in the earliest years. Connections are created between sounds, objects and emotions within the brain. In the absence of responsive caregiving the brain's architecture does not form as expected, which can lead to disparities in learning and behaviour (Centre on the Developing Child 2023). In Ireland the Department of Children and Youth Affairs has acknowledged the importance of the first five years of a child's development and created several social and education policies to increase the amount of time parents can spend with children, and to provide affordable and accessible early childhood education supporting learning through play.

The most recent neuroscience on brain research confirms the power of play for brain development. Play experiences contribute to the formation and strengthening of neural connections in the brain. Engaging in various activities during play stimulates different regions of the brain, contributing to cognitive, social, emotional and physical growth. For instance, play often involves scenarios that require problem-solving, critical thinking, and decision-making. Children learn to navigate challenges, make choices, and think creatively during play, fostering cognitive development. A number of theorists (including Piaget 1936; Vygotsky 1978) have particularly emphasised play as a mechanism for developing skills in symbolic representation and perspective-taking (Passer et al. 2009). Play is also an important mechanism for children's socio-emotional and cognitive development. The brain makes connections during play as children are given free rein to the imagination and consequently they can engage in new experiences, activities, roles and relationships on an on-going basis (Early Childhood Ireland 2023). Cooperative play allows children to understand and share the feelings of others. They learn to consider the perspectives of their playmates, fostering the development of empathy.

Laughter probably originated in play, which is important in the social and physical development of children even before they have the use of language. In the early years, eye contact, smiling, tickling, playing 'peek-a-boo' and laughter are the essential building blocks for future social engagement. A growing body of scientific literature suggests that laughter is good for the immune system and counteracts stress. As soon as we give way to laughter, electrical impulses are triggered by nerves in the brain, which set off chemical reactions there and in other parts of our body (Corballis 2014). Our endocrine system orders our brain to secrete its natural tranquilisers and painkillers, which ease anxiety and relieve pain. Laughing with someone else has double effects – it improves and consolidates interpersonal relationships.

Resource
The Centre for the Developing Child, Harvard University is a multidisciplinary team committed to driving science-based innovation in policy and practice. The Centre creates resources and provides information about child development, early childhood mental health, resilience, executive function and other core factors influencing the development of children facing adversity. Available from: <https://developingchild.harvard.edu/>

The Power of School Relationships

Child psychologist James Comer said no significant learning occurs without a significant relationship (2001). The quality of the teacher-student relationship is thus a key factor in academic success. Warm, caring, supportive student-teacher relationships, as well as other child–adult relationships, are linked to better school performance and engagement, greater emotional regulation, social competence and willingness to take on challenges (Osher et al. 2018). Teachers who establish positive, supportive and respectful relationships with their students create an environment conducive to learning, and have positive effects on student engagement, student motivation, student attendance, achievement, and test scores.

Positive learning environments and strong teacher student relationships can act as protective factors for children's social and academic development (Valiente et al. 2008). In identity formation, students need to feel like a part of the school or class group. Teachers are essential in creating a sense of belonging and connectedness within the learning environment. When students feel valued, respected and supported, they are more likely to engage actively in the learning process and feel a sense of ownership over their education. Trust is a crucial component of effective learning environments. When students trust their teachers and peers, they feel safe to take risks, ask questions and express their thoughts and opinions without fear of judgement.

A classroom in which all identities are respected promotes student achievement and attachments to school (Steele and Cohn-Vargas 2013), reinforcing their value and belonging. Teachers can create an identity-safe setting by cultivating diversity as a resource for teaching through regular use of materials, ideas and teaching activities that draw on referents to a wide range of cultures and exhibit high expectations for all students (Darling-Hammond et al. 2019).

School bonding (Sakiz et al. 2012) is another important factor influencing learners' progress in a formal learning environment. School bonding is similar to attachment in the way that it makes learners feel valued and secure, allowing them to explore new ideas. School bonding goes beyond learners' close relationships with teachers, and includes their relationships with peers. Students' sense of belonging involves a shared commitment with peers to succeed in school, to participate in extracurricular activities and to have a positive attitude towards learning (Kristic 2015). Peer relationships contribute to the social learning environment. Positive interactions with peers can enhance collaboration, teamwork and the development of interpersonal skills. Peer support also plays a role in reducing stress and anxiety. Collaborative learning, discussions and group activities provide opportunities for cognitive growth through the exchange of ideas and perspectives.

TED Talk

In this TED talk *Every Kid Needs a Champion*, educator Rita Pearson argues that every child deserves a champion, an adult who will never give up on them, who understands the power of connection. In New York in 2013, Pearson shares the secret to teaching students from disadvantaged backgrounds and discusses the value and importance of human connection.

Available from: <https://www.ted.com/talks/rita_pierson_every_kid_needs_a_champion?language=en>

Building School-Community Partnerships

Fundamentally learning happens in a context where wider social, cultural and structural factors are at play. To start we must acknowledge that relationships extend beyond the school community. In Bronfenbrenner's ecological systems theory, he locates the learner within a complex network of interactions between individuals and their environment, highlighting the influence of various systems on human development. Relationships, brain development, learning and growth should be understood within the wider context of these interconnected systems. School-community partnerships play a vital role in supporting child development by creating collaborative relationships between educational institutions and the broader community. These partnerships offer a range of benefits that contribute to the overall well-being and educational success of children.

Joyce Epstein's work has continually recognised the multifaceted nature of school-community partnerships and emphasises that successful involvement requires collaboration at various levels. Her work underscores the idea that schools, families and communities are interconnected and that positive outcomes for students are more likely when these entities work together. Research has shown that building strong relationships between the school, the family and the community increases academic outcomes for students, as the three groups work cooperatively with a shared responsibility to advance children's development (Willems and Gonzalez-DeHass 2012). In a series of meta-analyses examining the impact of parent involvement, Jeynes (2017) found consistent positive effects of parent involvement on academic achievement for children from pre-K through 12th grade.

For some learners, their access to nutrition, sleep, connection, relationships and safety (all of which influence learning) are determined by their socio-economic status, their home location, their family dynamics. By strengthening positive interactions with communities, families and peers, schools can 'foster environments rich in the developmental supports and opportunities needed to develop resilience in young people' (Worsley 2010: 6). Worsley has found in her work that these relationships significantly contribute to the success of students in school and beyond.

> ### Funded Research
>
> In the UK some of the context issues relating to learning were identified when the Education Endowment Foundation (a charity with the mission to improve the educational attainment of the poorest pupils in English schools) and The Wellcome Trust launched an 'Education and Neuroscience scheme' (2014). The aim of the initiative was to fund joint projects between educators and neuroscientists which would develop evidence-based interventions for use in the classroom. In total, 6 research projects were funded by the initiative and results can be found below:
>
> - SMART Spaces – testing the effect of a spaced learning approach on GCSE outcome. Available from: <https://educationendowmentfoundation.org.uk/projects-and-evaluation/projects/smart-spaces/>
> - Stop and Think: Learning Counterintuitive Concepts – A project to develop and test software that improves pupils' ability to 'inhibit' irrelevant prior knowledge when learning new concepts. Available from: <https://educationendowmentfoundation.org.uk/projects-and-evaluation/projects/learning-counterintuitive-concepts/>
> - Fit to Study: A programme to optimise the benefit of PE for brain function. Available from: <https://educationendowmentfoundation.org.uk/projects-and-evaluation/projects/fit-to-study/>
> - Teensleep: Testing the impact of a sleep education programme. Available from: <https://educationendowmentfoundation.org.uk/projects-and-evaluation/projects/teensleep/>
> - Sci-napse: Engaging the Brain's Reward System: Testing the impact of game-based rewards in secondary school science classes. Available from: <https://educationendowmentfoundation.org.uk/projects-and-evaluation/projects/engaging-the-brains-reward-system/>
> - GraphoGame Rime: Testing a computer programme designed to improve pupils' literacy through teaching phonics via 'rhyme analogy'. Available from: <https://educationendowmentfoundation.org.uk/projects-and-evaluation/projects/graphogame-rime>

The Effects of Social Isolation

Penal solitary confinement is illegal in many jurisdictions, precisely because observers have long recognised the damage caused by stripping away one of the most vital aspects of a human's life – interaction with others. To be lonely, as distinct from being alone, has been described as a silent jail, without cellmates, parole or bail. Loneliness is essentially the subjective feeling that we are lacking the social interactions and connections we need. It is a perception of a mismatch about what we expect versus our reality.

The American scientist, John Cacioppo, and his colleagues at the University of Chicago have researched in a decade-long study the effects of loneliness on health. They compared the health of two groups of adults in Chicago: a collection of people that identified themselves as chronically lonely, alongside those who believed themselves to be functioning, socially networked people. Drawing on research on brain imaging, analysis of blood pressure, immune response, stress hormones, lifestyle patterns and gene expressions, the study revealed just how interdependent human beings are. Loneliness contributes to physical ill health. 'And so, no, it may not kill us immediately, but it will nevertheless hasten our demise' (Cacioppo 2009).

There are two general kinds of loneliness. The first is transient, the kind we are all prone to feel from time to time, for example, when we move to a new location for work, children leave the family home or we endure a major social disruption to the routines of our lives. But this kind is largely fleeting. We mostly succeed in establishing new relationships, and so the initial sense of painful disruption ultimately recedes. So transient situational loneliness doesn't have huge long-term consequences for health.

The second kind of loneliness is termed persistent loneliness. According to the *Campaign to End Loneliness (n.d.)*, research suggests that one in ten GP visits today in England are being booked simply because people crave the social interaction they currently lack (Duerden 2018). In Japan, over half a million teenagers and young people are increasingly living in an exclusively virtual world, interacting with friends online and not leaving their home or physically meeting with others for months at a time. They even have a term for this phenomenon in Japan – *hikikomori* – to describe these young people.

Taking into account pre-existing health conditions, socioeconomic status, smoking, alcohol consumption, obesity, race, life satisfaction, physical activity and use of preventive health service, one study examining the people of Alameda County, California, found that in every age and sex category, people with the fewest social ties were 3 times more likely to die over a 9-year period than those who reported the most social ties (Gratton and Scott 2016).

Technology has made our lives simpler while at the same time changed the very concept of human interaction as we know it. This is problematic, and the problem is progressively global. In 2013, *The New York Times* August magazine reported that 1 in 3 Americans over the age of 45 identifies as 'chronically lonely', while the *British Red Cross* reported in 2017 that over 9 million people in the UK across all adult ages are either often or always lonely.

A study conducted by Bonsaksen et al. (2023) in 4 countries during COVID-19 found that high levels of social media use related not only to loneliness, but to mental distress more generally. More time spent digesting other people's happiness on social media may accelerate one's own feelings of loneliness and distress, possibly fuelled by envy, as shown in a recent study (Wang et al. 2020). Furthermore, people may experience social media as a poor substitute for face-to-face contact, as virtual contact may not feel as real or as meaningful, as it would in real life (Yao and Zhong 2014). The campaign to end loneliness in the UK reported that people aged 30–45 were expressing high or chronic levels of loneliness when compared with those aged between 16 and 29 and over 2 times more likely to report feeling lonely often or always than those over 70. Social comparison and wider expectations about how young adults should live their lives also contribute to loneliness.

Loneliness has been found to increase the risk of mental health problems such as depression, anxiety and stress (Richardson 2017). It is a common metaphor to speak of social exclusion or loneliness as being 'painful'. This turns out to be true. Researchers have found that the area of the brain activated during physical pain is activated more often during times of separation and isolation. It appears that the brain shows a faster rate of decline in neural connections during times of loneliness (Eisenberger et al. 2003). Not only are these things not good for a person's mental health, but they also put people at risk of social withdrawal, compounded by the fact that inflammation leaves some people less motivated to seek out the company of others. All of which can create a vicious spiral of loneliness that may be hard to break. It is as if lonely people are constantly on the lookout for threats, especially in a social context. This bias towards the negative, also affects how they view relationships, making them more likely to adopt protective behaviours that mean they are less likely to put themselves out there, and more prone to expect rejection.

As we will see in later chapters, loneliness often causes sleep issues, too, which can also be explain this hyper-vigilance stopping people from switching off. And possibly related to sleep issues, loneliness is linked to impaired executive function, accelerated cognitive decline and progression towards Alzheimer's disease (Cacioppo et al. 2009).

Conclusion

In a very practical sense social engagement challenges us to communicate frequently and to participate in complex social interactions such as listening, story-telling and showing empathy. It fosters a dynamic environment for which mobilisation is necessary. It requires a commitment to family, friends and community groups that may promote purpose and role in terms of membership of different social activity support networks, of assistance to others. Social networks fostered by social engagement can assist in both giving and receiving support (de Lange 2022). In terms of learning, social engagement contributes to psychological empowerment gained from a sense of belonging and self-esteem and the feeling that we 'matter' to others.

The benefits of social communication are compelling – those with strong relationships appear to have less cognitive decline and live longer and more active lives. Studies show that people with larger social networks have a larger amygdala, the brain region used most in social interaction (Banks 2015). As we move through the life-cycle we go through a series of transitions. Some transitions are expected and anticipated while others may be caused by unexpected life events. Whatever the cause we can recognise a time of transition because it triggers a change in 4 key areas in our life. These include a change in our roles, relationships, routines and assumptions. It is during these times our brain is more open to change and friendships are needed in adapting successfully (Chickering and Reisser 1993).

Part of staying involved in society and maintaining a complex and novel environment for our brains to thrive is sustaining a network of social connections. Humans require love and need to express love. The development of strong family and friendships is not a given, nor is it always easy. However as part of a lifestyle for a healthy brain the importance of proactive nurturing of relationships cannot be underestimated. As the research above demonstrates the skills in communicating with others begin in childhood and continues through to old age. As we have mentioned we are social animals by nature and we have evolved to be part of a community made up of interconnected social networks. It makes sense that our social life, with rich and meaningful connections, is so good for the brain.

References

Banks, A. (2015). *Wired to Connect: The Surprising Link Between Brain Science and Strong, Healthy Relationships.* New York: Penguin Random House LLC.

Bloom, F., Beal, F. M. and Kulfer, D. (Eds). (2003). *The Dana Guide to Brain Health.* New York: The Free Press.

Bonsaksen, T., Ruffolo, M., Price, D., Leung, J., Thygesen, H., Lamph, G., Kabelenga, I. and Geirdal, A. Ø. (2013). Associations between Social Media Use and Loneliness in a Cross-national Population: Do Motives for Social Media Use Matter? *Health Psychology and Behavioral Medicine*, 11(1):2158089, pp.1-18.

Bowlby, J. (1969). *Attachment and Loss.* New York: Basic Books.

Butler, G., Grey, N. and Hope, T. (2018). *Manage Your Mind.* Oxford: Oxford University Press.

Cacioppo, J. and Patrick, W. (2009). *Loneliness: Human Nature and the Need for Social Connection.* W. W. Norton & Co.

Campaign to End Loneliness <https://www.campaigntoendloneliness.org/press-release/younger-brits-report-higher-levels-of-loneliness/>

Chickering, W. and Reisser, L. (1993). *Education and Identity.* San Francisco: Jossey-Bass Publishers.

Corballis, M. (2014). *A Very Short Tour of the Mind: 21 Short Walks around the Human Brain.* London: Duckworth Overlook.

Cozolino, L. (2014). *The Neuroscience of Human Relationships.* New York: W. W. Norton.

Darling-Hammond, L., Flook, L., Cook-Harvey, C., Barron, B. and Osher, D. (2019). Implications for Educational Practice of the Science of Learning and Development. *Applied Developmental Science*, 24(2), 97–140.

De Lange, C. (2022) *Brain Power: Everything You Needed to Know for a Healthy Happy Brain.* London: Michael O'Mara Books.

Duerden, N. (2018). *A Life Less Lonely: What We Can All Do to Lead More Connected, Kinder Lives.* Green Trees: Bloomsbury Publishing Plc.

Dunbar, R. (2011). *How Many Friends Does One Person Need? Dunbar's Number and Other Evolutionary Quirks.* London: Faber & Faber.

Eisenberger, N., Lieberman, M. and Williams, K. (2003). Does Rejection Hurt? An fMRI Study of Social Exclusion. *Science*, 302(5643), 290–292.

Epstein, J. L., Sanders, M. G., Simon, B. S., Salinas, K. Clark, Jansorn, N. Rodriguez and Van Voorhis, Frances L. (2002). School, Family, and Community Partnerships: Your Handbook for Action (2nd edn). Thousand Oaks, CA: Corwin Press.

Flanagan, C. (2021). AQA Psychology For A Level & AS - Your Guide To Exam Success. London: Illuminate Publishing.

Gallese, V., Fadiga, L. and Fogassi, L. (1996). Action Recognition in the Premotor Cortex. *Brain*, 11.9(2), 593–609.

Gawande, A. (2015). *Being Mortal: Illness, Medicine and What Matters in the End.* New York: Profile Books.

Giddens, A. (1997). *Sociology* (3rd edn). Cambridge: Polity Press.

Gratton, L. and Scott, A. (2016). *The 100 Year Life: Living and Working in an Age of Longevity.* London: Bloomsbury.

Guillebeau, C. (2014). *The Happiness of Pursuit.* New York: Harmony Books.

Harrold, G. (2009). *Look Younger, Live Longer: Slow the Ageing Process.* London: Orion Books.

Kenyon, T. (2015). *Brain States.* Litaia Springs, GA: World Tree Press.

Kosslyn, S. and Rosenberg, R. (2001). *Psychology: The Brain, the Person, the World.* Boston: Allyn and Bacon.

Lawton, G. and Webb, J. (2017). *How to Be Human: The Ultimate Guide to Your Amazing Existence.* London: New Scientist.

Macknik, S. and Martinez-Conde, S. (2011). *Sleights of Mind: What the Neuroscience of Magic Reveals about Our Brains.* London: Profile Books.

Norton Duerden, N. (2018). *A Life Less Lonely: What We Can All Do to Lead More Connected, Kinder Lives.* Green Trees: Bloomsbury Publishing Plc.

Nussbaum, P. (2003). *Brain Health and Wellness.* Tarentum, Pennsylvania: Word Association Publishers.

Richardson, T., et al. (2017). The Relationship between Loneliness and Mental Health in Students. *Journal of Public Mental Health*, 16(2), 48-54.

Shonkoff, J. P. (2017). Breakthrough Impacts: What Science Tells Us about Supporting Early Childhood Development. *Young Children*, 72(2), 8–16.

Strauch, B. (2011). *The Secret Life of the Grown-Up Brain: Discover the Powerful Talents of the Middle-aged Mind.* London: Penguin Group.

Vygotsky, L. (1978). *Mind and Society: The Development of Higher Mental Processes.* Cambridge, MA: Harvard University Press.

Wang, W., Wang, M., Hu, Q., Wang, P., Lei, L. and Jiang, S. (2020). Upward Social Comparison on Mobile Social Media and Depression: The Mediating Role of Envy and the Moderating Role of Marital Quality. *Journal of Affective Disorders*, 270, 143–149.

Yao, M. Z. and Zhong, Z.-j. (2014). Loneliness, Social Contacts and Internet Addiction: A Cross-lagged Panel Study. *Computers in Human Behavior*, 30, 164–170.

CHAPTER 10

Food for Thought

> 'Let food be your medicine and medicine be your food'.
>
> (Hippocrates, cited in Nourse 1969: 11)

Nutrition is one of the most powerful lifestyle influences on learning and education. As a general rule, good nutrition for the body is good nutrition for the brain. Food we eat affects the way we feel and our ability to enjoy life. We all intuitively appreciate that the foods we eat shape our thoughts, actions, emotions and behaviour. When we eat too much, we feel sluggish; when we are tired, we crave coffee; when we feel low, we reach for chocolate.

Food and drink provide the energy and nutrients we need for breathing, developing and thinking. We need a regular supply of energy to power our heart so that it can pump blood carrying oxygen and nutrients to our brain. Unlike other organs, the brain does not store energy. In order to function it needs a constant supply of two major fuels: oxygen and glucose. As we will see later in this chapter the bacteria in our digestive system also plays a critical role in this process (Anderson et al. 2017).

As the human body changes throughout the life cycle, so too do our nutritional needs vary and change. During some stages of our lives, and in certain circumstances, our body and brain require extra energy to cope with the additional demands that are made. These include pregnancy, the early years of growth, adolescent development, times of illness, episodes of chronic stress, performing strenuous physical or intellectual activities, and so on. There is now strong evidence to suggest that as we age, our diet has a huge impact on our gut which in turn, influences our brain and the prevention or delay of many of its degenerative diseases.

In this chapter we will address food from the perspective of learning for brain health, keeping in mind the important and pervasive role that it plays in our lifestyle choices, and the impact that our lifestyle choices have in turn on how we perform. We suggest that small, thoughtful, sensible changes in what we eat each day can gradually produce significant benefits, and can lead to rapid improvements in our learning capacity and well-being. The first step on this journey is to understand the role of food in our lives and the positive or negative impacts it has, not only on our physical health, but more importantly on cognitive enhancement.

The Greedy Brain

The brain is a high-energy organ. Despite making up only 3 per cent of our body weight, it consumes up to 20 per cent of our total energy supply. Of all the organs in the body, the brain is the greediest in its fuel consumption (Perlmutter 2016). Its fuel is a type of sugar called glucose. It is the main fuel used by all the body's cells, but unlike other cells, brain cells cannot turn to other types of fuel – neurons depend exclusively upon glucose. The only exception is in rare cases where a ketogenic diet over time results in neurons switching their metabolism to fat rather than glucose. Overall, the brain burns oxygen and glucose at ten times the rate of all other body tissues. In fact, it uses up so much energy that it dies if deprived of oxygen for only a few minutes. Even activities such as thinking and sleeping require significant amounts of energy.

There is a distinction to be made between food and nutrition. Food is something that provides nutrients for the body, which are essential for life. Nutrients are the substances that provide energy for all our activity and growth, as well as all the important functions of the body that allow us to breathe, stay warm, move and think and do unconscious activities. Nutrients are components of food needed in adequate amounts for proper growth, reproduction and the leading of normal lives. They also repair the body and brain and help keep the immune system healthy.

Nutrition, on the other hand, is food at work in the body. It includes everything that happens when you eat food that is used by the cells and neurons to make all the contributions that our body needs. Nutrition is brain fuel that enables the brain to work. As noted, typically our daily food consumption supplies the brain with glucose for fuel. As we will see, not all nutrients are acquired from food. For example, Vitamin D from the sun contributes to the maintenance of bones, teeth and the normal functioning of the immune system.

All foods contain two main categories of nutrients: macronutrients and micronutrients. Macronutrients are required in large amounts for healthy growth and development, forming the basis of every diet and providing energy for all the body's everyday functions and activities. Vitamins and minerals make up the micronutrients, so called because they are found in tiny amounts in foods. Unlike macronutrients, vitamins and minerals do not provide energy and are needed in small amounts, but they provide a critical role in the normal functioning of the body and digestive processes to ensure good health.

The brain needs to balance the energy we take in from the food we eat with the energy we expend in the course of our daily lives. A person's energy requirements depend on various factors, including age, gender, muscle mass, physical activity, body temperature, and whether they are still growing. Excess food intake is stored in the body as fat leading to weight gain. Of course, the reverse is also true: if we expend more energy than we take in, we may lose weight. In extreme cases both can lead to medical conditions.

Meet the Second Brain

Conventional wisdom teaches that anxiety and depression contribute to physical problems in the body, but what if it is the other way around? Recent research has shown that the root of many brain-related disorders could be not within the brain but rather be in the body, especially within the gut (Roshini 2022). The expression 'gut feeling' is an accurate expression of how we feel about a situation, as we now know that the gut is closely involved with mood. Furthermore, the idea of altering behaviour and mood by changing the contents of the gut is very new.

Thanks to breakthroughs in sophisticated technology, neuroscientists have demonstrated the intimate, two-way communication between the brain and the digestive system. The person most associated with promoting the idea that there is a strong link between the health of our gut and our brain health is Michael Gershon (1998), who called the gut *The Second Brain*. He used the term to refer to the network of nerves surrounding our gut. More recent scientists like Anderson, Cryan and Dinan (2017) have focused on the complexity of having a healthy microbiome and the relationship between mood, food and the new science of psycho-biotics of the gut-brain connection.

The brain communicates through the central nervous system, while the gut has its own enteric nervous system. The enteric nervous system is an interconnected system of neurons that controls the gastrointestinal system and is often referred to as the second brain. This system consists of the nerve cells embedded in the walls of the intestines. It controls the movement of food through the gut. This process is always going on, and despite its autonomy, the enteric nervous system maintains a constant interaction or conversation with our actual first brain.

A large part of that conversation concerns the health of our gut and that of our brain. How does this conversation happen? The gut and brains are connected by the vagus nerve. This particular nerve is well named because the word '*vag*' means vagrant or wandering. This nerve wanders through the diaphragm, between the lungs and the heart, up along the oesophagus, through the neck to the brain, via the brain stem, hypothalamus and cortex. It uses acetylcholine as a neurotransmitter and its primary purpose is to communicate with the brain. It is also the fastest route by which the gut sends information to the brain.

The purpose of this gut-to-brain signalling means that the brain receives information about what is going on in the gut. Likewise the brain, through the central nervous system sends information back to the gut. So this nerve works something like an old-fashioned telephone switchboard at a large organisation transferring messages to and from many different units or as a 'walkie-talkie' is used in a modern security system. A healthy gut does not transmit minor, unimportant digestive signals to the brain.

It only signals the brain when something is not right or uncomfortable. Likewise the brain only signals the gut when something is not right.

Stress is thought to be among the most important 'topics' in the gut-brain communication axis. When the brain senses a possible stressful event (e.g. an upcoming exam or presentation), it naturally wants to deal with it. To do this it needs energy, which it borrows mainly from the gut. The gut is informed of the emergency situation, via the vagus nerve, and is instructed to obey the brain in this exceptional situation. The gut in turn is kind enough to save energy on digestion, thus producing less mucus and reducing the blood supply. This negative stimulus can cause loss of appetite or diarrhoea. Such a situation as described is short-lived and harmony is quickly restored.

If stress continues for a long time and becomes chronic, it can seriously damage how the gut functions, with serious health implications. Ideally the reciprocal chemistry between the 'upper brain' and 'lower brain' must work in harmony or there will be 'confusion' in the gut and 'misery' in the head. It should come as no surprise that improving the relationship between our gut and our brain can improve overall health, sense of mental wellness and academic performance. Optimising our gut health thus means looking after the bacteria in our digestive system, particularly maintaining the function of the gut-barrier – the wall that separates the interior of our gut and our bloodstream.

Expert Insight

Professor Usha Goswami, University of Cambridge

In conversation with the authors, Prof Usha Goswami, Director of the Centre for Neuroscience in Education at the University of Cambridge, highlighted some of the key learnings' links between physiological health and learning.

Diet affects our ability to learn

One of the really key findings in the last decade has been that we have brain cells in our gut, and this brain-gut axis is fundamental. This axis ranges from the gut-brain reactions that children have which can then impede their learning in the classroom, to the very simple parameters of how good children's diet is for children worldwide.

Does one's diet foster a healthy gut? We already know that if we're not getting enough nutrients, or we don't have a good diet, like in some parts of the developing world, we don't benefit as much from education because we just haven't got the nutrition that your brain needs to learn. Even in Westernised societies, this has an impact because the kind of foods you're being given to eat are going to affect how your gut responds to situations in educational settings, and how that can either impair or enhance your ability to learn in those settings.

Phase relations are fundamental to language development

There are very intricate rhythmic relations in all levels of how the brain works, not just the electrical firing itself, but the signalling that follows from that firing. The same things actually happen in your gut. These phase relations, which we think of as rhythmic timing relations are absolutely fundamental to individual differences in people. The work I'm doing on language, for example, shows that phase relations actually make a big difference over the learning trajectory. What neuroscience suggests is the elements of the timing patterns in the brain that are taking in that acoustic, sensory information – when they're slightly out of kilter, that's when a child is at risk for a language learning disability. Teachers are very powerful agents in the learning of children. The simple message for a preschool or primary school teacher would be that none of the activities centred around rhythm or rhyme are a waste of time – they are absolutely fundamental to language development.

The motor system is involved in everything we learn

Even if you're reading words like kick versus lick, your motor cortex will be activated by the bit for your tongue when you read the word lick, then a bit for your foot when you read the word kick. All of our neural systems are integrated, and moving around your environment as a toddler or doing things with your hands in primary school – these are really fundamental to learning. It is important for learners to do things themselves. Neuroscience massively supports the idea that teaching should be multi-modal.

Recent Research
Goswami, U. (2022). Language Acquisition and Speech Rhythm Patterns: An Auditory Neuroscience Perspective. *Royal Society Open Science*, 9(7), 1–14.
Goswami, U. (2019). *Cognitive Development and Cognitive Neuroscience: The Learning Brain* (2nd edn). Routledge.
Power, A. J., Mead, N., Barnes, L. and Goswami, U. (2013). Neural Entrainment to Rhythmic Speech in Children with Developmental Dyslexia. *Frontiers in Human Neuroscience*, 7(777), 1–19.

Brain Supporting Foods

Our gut is an incredible, dynamic, self-controlling system. It has built-in systems of checks and balances to keep the digestive process on an even keel. Cellular transactions are happening every second, without our even knowing it, to help maintain our digestive system's overall balance and preferred settings – what scientists call homeostasis. A healthy gut is similar to an ecological system; to thrive, it requires a wide diversity of bacteria competing with each other. This is why neuroscientists recommend eating different foods of different colours. One of the reasons is that this variety is good for the bugs that live in our gut, and their associated genes, collectively known as our microbiome (Anderson et al. 2017).

Bacteria are single-celled tiny organisms that live in soil, water and air and also inhabit the body. Their shapes vary and include oval, spherical, rod-like and spiral. Flagella (whip-like strands) help bacteria to move. They are amazingly adaptable and versatile: they grow, reproduce, interbreed, mutate, swap genes, multiply, produce vitamins and seem to have incredible powers over one's health. The community of microbes living in our gut is like another organ in our body (Enders 2016). When everything is running smoothly, we pay no attention to our gut. It is best left on autopilot, until something goes wrong.

There are about 100 trillion bacteria that live both on you and in you, with most of them living in your gut. The highest concentration of bacteria is found where the digestive process is almost finished – in the colon. Of the more than 1,000 species that commonly live in and on the human body, each of us harbours only around forty species in our gut, with each having particular functions. Most of them are beneficial. It is interesting to note that one aspect of our uniqueness is that everyone's bacterial population is different and its composition changes after every meal – our unique bacterial fingerprint!

We have evolved a complex and mutually beneficial relationship with gut bacteria. Our side of the bargain is straightforward: we provide them with a cosy and warm environment in which to live, and we feed them. The great news is that we only feed them our leftovers – stuff we can't use. Our diet influences the colonisation of our gut (Perlmutter 2016). Most of the time beneficial bacteria thrive, but if we are ill, stressed or taking antibiotics, our gut flora suffer. Our task, therefore, is to keep the beneficial bacteria healthy and happy and to make sure they outnumber the less helpful ones.

We need food when the brain triggers hunger sensations caused by contractions in the stomach. Most of the time however, internal mechanisms in the brain control our eating habits such as when we eat, what we eat and how often we eat throughout a 24-hour cycle. These mechanisms can be either positive or negative and are influenced by two important factors: our habitual level of physical activity and our mental association of eating within different social and environmental situations. Both these factors are immediately relevant to brain health and learning.

Food plays a critical role in the battle between beneficial and harmful bacteria in the digestive tract. This is why competing bacteria are good for the gut. This is often referred to as the hygiene hypothesis, which argues that exposure to a broad range of microbes throughout life helps to develop robust immune defences (Pallardy 2006). However the aim of gut hygiene is not to eliminate bacteria but to feed it. Even harmful bacteria can be good for us when the immune system uses them to build up defences; for instance, salmonella is only dangerous when it turns up in greater numbers.

We are advised to eat as wide a variety of foods as possible, to eat appropriate food only and to make sure we get at least some food from each of four groups each day. We have probably seen the food pyramid guide which breaks food up into four types or groups. Usually the pyramid includes the bread cereal group at the base, followed by the vegetable and fruit group, and the fish and meat group with the fat and sugar group taking up the smallest section of the pyramid at the top.

Food for Thought 173

Figure 15: Food for Brain Health – Image by svtdesign/<stock.adobe.com>

What we eat and how we eat determines a healthy balance between competing bacteria in our gut and functioning neurons in our brain. The food choices we make are within our control. Some medical research on the gut microbiome (Wang et al. 2021) suggests that a diet rich in fibre and probiotics found in berries, fruits and nuts act as antioxidants and are thus considered beneficial food for the brain.

Antioxidants for Brain Health

The most common example used to explain antioxidants is to imagine what would happen if you cut an apple in half and allowed it to sit on the counter. In a short time the sliced portion exposed to the air would turn brown or oxidise. If you took another apple, cut it in half, but this time squeeze some lemon juice on the exposed surfaces, it would remain fresh looking without much browning or oxidation for a much longer period. The antioxidants contained in the lemon juice serve as natural protective preservatives for the apple tissue. Similarly, antioxidant found in certain foods act to guard body tissues and our brain from dangerous damage.

Including foods in our diet that contain specific brain-building nutrients is critical in facilitating our best mental performance and delaying cognitive impairment. We can start protecting our brain by including foods high in antioxidants in the food choices we make. It is important to remember that the brain works best with we eat a variety of foods from the basic 'pyramid' groups. If a healthy balanced diet seems too hard to maintain, then follow the 80/20 rule. If 80 per cent of the time you are eating the best healthy food, you can afford the odd indulgence 20 per cent of the time.

In many countries today, average diet can actually be toxic and a form of malnutrition. Giulia Enders in her book, *Gut: The Inside Story of Our Body's Under-Rated Organ,* writes that some 80 per cent of the processed foods found on the shelves of modern-day American supermarkets contain added sugar. Studies show that a high-fat, low-fibre diet increases illness-inducing chemicals and a high-fat high-sugar diet can degrade the blood brain barrier, allowing dangerous toxins to access our brain. Eating a variety of different fruits provides ample antioxidants to prevent such damage – some of the most popular include oranges, watermelons, apples, mangos, apricots, plums, dates, grapefruits, avocados and grapes.

By actively seeking out and incorporating antioxidants that specifically contain valuable components, we ensure that our brain is provided with regular sources of energy, nourishment and re-generation. Antioxidants are primarily found in fruits and vegetables that are rich in vitamin A, vitamin C, vitamin E and vitamin D. The body cannot manufacture these nutrients so they must be supplied in the diet. Certain antioxidants foods have been shown, from decades of research, to slow down damage to the brain. A combination of antioxidants is brain beneficial as they all have different protective roles.

Fruits: The antioxidants found in fruits have been positively linked to increased mental function. The true value in consuming a variety of fruits is that most contain generous amounts of Vitamin C, a powerful antioxidant.

However how Vitamin C reaches the brain is an interesting process. The brain has a spectacular defence network known as the blood brain barrier (BBB). The barrier's job is to keep unwelcome molecules outside to protect the most precious organ in the body. The brain is quite clever when it comes to absorbing Vitamin C, as it permits an altered version only.

Berries contain high levels of antioxidants and within the neuroscience community are often referred to as brain berries. Just adding these to your diet can delay mental decline by two and a half years or more (Devore et al. 2012). They contain nutrients that can enhance neuron communication within the brain and also help to repair damaged neurons (Amen 2015). In 1999, researchers at the Human Nutrition Research Centre in Boston announced that they had significantly improved rats' brainpower by feeding them blueberries (one reason rats are most frequently used as experimental animals is that their nervous system is in many basic ways similar to the human nervous system). Since then, thousands of laboratory rats have been given blueberries in an attempt to find our more. In one study, rats which were fed blueberries developed better balance and coordination as they aged compared to their study counterparts. In another study, rats fed a blueberry-enriched diet that were then given a stroke had a quicker recovery time and only lost 17 per cent of neurons in the hippocampus, compared with 43 per cent neuron loss in rats not using blueberries (Gomiz-Pinilla 2008). These studies point to the potential reversal of some age-related impairment in both movement coordination and memory loss in humans.

The more vibrant the colour of the berry, the more health promoting nutrients our brain receives. Dark tinted berries contain a certain antioxidant called *anthocyanins* are beneficial plant pigments that give berries their deep red, purple or blue hues. We recommend to 'taste the rainbow' through blueberries, blackberries, cranberries, gooseberries, raspberries and strawberries in order to obtain a wide variety of antioxidants to nourish and protect the brain.

Nuts and seeds: Nuts and seeds are inexpensive and are tiny nutritional gems. They are valuable sources of protein; nuts are dried tree fruits and are contained within hard shells, while seeds are the embryo and food supply of new plants. Both are packed with proteins and a good source of vitamin E. Regular consumption of nuts and seeds appears to delay cognitive ageing.

Studies confirm that older people who eat a variety of nuts on a regular basis had the intellectual powers of people five to eight years younger. One possible explanation for this is that nuts are rich in an amino acid that relaxes blood vessels and eases blood flow to the brain (Amen 2005). Nuts and seeds in the food we eat affect the neurotransmitters, such as serotonin, dopamine and norepinephrine, which carry messages from one neuron to another and affect mood as well as thoughts and actions.

Proteins: Brain cells need proper nutrition to carry on their functions just like any other cell in the body. The myelin sheath covering neurons in the brain acts like insulation covering electrical wires. It speeds transmission of electrical signals to the brain. Deficiencies in proteins delay nerve-impulse transmission impacting brain function. Key protein foods include meat, fish, eggs and dairy products such as milk and cheese. Sources of Omega-3 are particularly important for the efficient functioning of neurons and glial cells in the brain. It is found in oily fish (mackerel, herrings, sardines and kippers), as well as in Brazil nuts, hazelnuts, almonds, and sesame and sunflower seeds.

Sunlight and Vitamin D: Vitamin D is hard to find in sufficient quantities in food, even though dairy products such as eggs and milk contain small amounts. The main natural source of vitamin D is from sunlight. When sunlight hits our skin, a chemical in our skin absorbs some of the light and produces vitamin D. Information published by the Commission the European Union suggests that 10–15 minutes exposure to sunlight leads to the production of 10,000 IU (International Units) of vitamin D. We may not think that sunlight is something that affects your brain health. The benefits of sunlight include providing vitamin D, which in turn kills germs, strengthens the immune system and improves quality of sleep (Agus 2016). This essential nutrient controls body hormones and cell growth, it helps our body absorb and use the calcium it needs for strong bones and teeth and it works to develop a healthy immune and nervous system.

Food, Mood and Learning

To maintain an even supply of glucose to the brain, breakfast is considered the most important meal of the day and it's thought to be beneficial for cognitive and academic performance across all student age groups. Having a nutritious breakfast positively influences various aspects of cognitive function, mood and behaviour. A nutritious breakfast helps stabilise blood sugar levels, providing a steady supply of glucose to the brain. This contributes to sustained attention and concentration throughout the morning, reducing the likelihood of fatigue and distraction in the classroom.

There is now a growing body of scientific evidence to suggest that diet can play a significant role in improving academic performance in children and adolescents as well as displaying fewer behavioural problems. The link between diet and academic achievement shows that various dietary components such as iron and Omega 3 have essential roles in brain development and functioning. Furthermore the brain requires significant and regular amounts of energy to function optimally. Regular eating throughout the day provides an opportunity for delivery of nutrient rich food, linked with high intake of essential micronutrients. This has led many countries, including Ireland, to have a national school breakfast programme.

Children and adolescents may be particularly sensitive to the nutritional benefits of breakfast on brain activity, as they have a higher brain glucose metabolism than do adults (Chugani 1998). In 2009, Adolphus et al., conducted the first systematic review of all the evidence examining the effect of breakfast on cognitive function (measured using tests of attention, memory, reaction time and executive function). The most consistent support for the benefit of breakfast was for attention, memory and executive function. In a follow up study in 2016, Adolphus et al., found that tasks requiring attention, executive function and memory were facilitated more reliably by learners who had consumed breakfast, when compared to learners who had fasted.

To examine the effects of nutrition on academic performance among university students, Tracey Burrows and her colleagues carried out a systematic review of the scientific literature on the association between dietary intake and academic achievement in college students. They reported that 'given the differences in the learning environment and learning styles in schools compared with university, as well as the inherent differences in the life stages of students, it is notable that the findings in the two educational settings are similar. In the current review, the positive effect of breakfast consumption in academic achievement was demonstrated' (2016: 9–10).

Conclusion

We are all familiar with the advice to eat the recommended daily consumption of the five to six different food groups as commonly displayed in the nutrition food pyramid. This dimension of food is beyond the scope of this chapter and only briefly mentioned above. This chapter focused on the importance of what we eat for the benefit of our brain. If there is one thing neuroscience tells us is that there is no specific food for brain health, but rather suggests it is better to apply the rainbow principle – namely ideally eat foods of many different colours and the darker the better (Chatterjee 2018).

Scientists have only recently begun focusing on the importance of having a healthy microbiome for our overall mental and physical well-being. An ideal microbiome is a diverse one, capable of adaptation with different bacteria cooperating together. Think of our own gut bugs like staff in our digestive factory that are producing the products we need to stay alive (Nerurkar 2024). To maintain optimal health for learning we need to ensure that all departments are adequately staffed and the teams are proportionate. Our microbiome represents a key component of our body's defence system against the outside world. The food we eat and its effect on our gut bacteria is intimately linked with the activity in our body's immune system, all the more so, as the majority of our immune system is found in and around the gut. Because the body is so interconnected, by feeding the microbiome with rainbow-coloured foods, we are also strengthening other parts of it, especially the brain.

References

Adolphus, K., Lawton, C. L., Champ, C. L. and Dye, L. (2016). The Effects of Breakfast and Breakfast Composition on Cognition in Children and Adolescents: A Systematic Review. *Advances in Nutrition*, 7(3): 590S–612S.

Agus, D. B. (2016). *The Lucky Years: How to Thrive in a Brave New World of Health.* London: Simon and Schuster.

Amen, D. G. (2005). *Make a Good Brain Great.* New York: Harmony Books.

Anderson, S., Cryan, J. and Dinan, T. (2017). *The Psychobiotic Revolution: Mood, Food, and the New Science of the Gut-Brain Connection.* Washington: National Geographic.

Burrows, T., Whatnall, M., Patterson, A. and Hutchesson, M. (2016). The Association between Dietary Intake and Academic Achievement in College Students: A Systematic Review. *Healthcare (Basel)*, 5(4), 60.

Chatterjee, R. (2018). *The 4 Pillar Plan: How to Relax, Eat, Move, Sleep Your Way to a Longer, Healthier Life.* London: Penguin Random House.

Chugani, H. T. (1998). A Critical Period of Brain Development: Studies of Cerebral Glucose Utilization with PET. *Preview of Medicine*, 27, 184–188.

Devore, E. E., Kang, J. H., Breteler, M. M. and Grodstein, F. (2012). Dietary Intakes of Berries and Flavonoids in Relation to Cognitive Decline. *Annals of Neurology*, 72(1), 135–143.

Enders, G. (2016). *Gut: The Inside Story of Our Body's Most Under-rated Organ.* London: Scribe.

Gershon, M. (1998). *The Second Brain: The Scientific Basis of Gut Instinct and a Ground Breaking New Understanding of Nervous Disorders of the Stomach and Intestines.* New York: Harper.

Gomez-Pinilla, F. (2008). The Effects of Nutrition and Brain Function. *Nature Review Neuroscience*, 9, 568–578.

Nerurkar, A. (2024). *The 5 Resets: Rewire Your Brain and Body for Less Stress and More Resilience.* London: Harper-Collins Publishers.

Nourse, A. E. (1969). The Hippocratic Corpus. In *The Body.* Netherlands: Time-Life International.

Pallardy, P. (2006). *Gut Instinct: What the Stomach Is Trying to Tell You.* Paris: Rodale.

Perlmutter, D. (2015). *Brain Maker: The Power of Gut Microbes to Heal and Protect Your Brain – for Life.* London: Yellow kite.

Perlmutter, D. and Loberg, K. (2016). *The Grain Brain: Whole Life Plan.* London: Little, Brown and Company.

Roshini, R. (2022). *Gut Renovation: Unlock the Age-defying Power of the Microbiome to Remodel Your Health from the Inside Out.* London: William Morrow.

Wang, D. D., Nguyen, L. H., Li, Y., Yan, Y., Ma, W., Rinott, E., Ivey, K. L., Shai, I., Willett, W. C., Hu, F. B., Rimm, E. B., Stampfer, M. J., Chan, A. T. and Huttenhower, C. (2021). The Gut Microbiome Modulates the Protective Association between a Mediterranean Diet and Cardiometabolic Disease Risk. *Nature Medicine*, 27(2), 333–343.

CHAPTER 11

Exercise and Neurogenesis

'Exercise has emerged as the closest thing we have to a magic wand for the brain'.

(Strauch 2011: 125)

Physical exercise serves as both healer and fuel for both the body and brain. If we had to choose just one thing to keep our brain working to the best of its abilities, we would suggest you choose exercise. Very few lifestyle activities, with the exception of exercise can both prevent and help in the treatment of so many modern diseases like type 2 diabetes, obesity, depression and Alzheimer's disease. In this context perhaps the most important insight about the benefit of exercise is the growing awareness of maintaining a healthy and well-functioning brain. Keeping active helps young people and adults alike with learning, memory, focus and creativity as we will come to learn in this chapter.

The wide-ranging consequences of sitting too much and moving too little are not new. It is no secret that our society has shifted to a more sedentary lifestyle, where many of us spend a large portion of our days sitting or being largely inactive in some way. Being inactive is not in itself harmful, but for prolonged periods it can negatively influence the body. According to the World Health Organisation (WHO), inactivity is tied to more than five million deaths worldwide each year and is the fourth most common preventable cause of death behind smoking, hypertension and obesity.

Exercise affects the brain in many ways. It strengthens the heart and improves blood circulation, as well as providing essential nutrients to the brain. The increased blood flow raises the oxygen levels stimulating the nervous system, increasing the metabolic rate in order to better fight disease. In fact, today's most exciting research in exercise science involves deciphering the ways in which activity affects the learning brain. Recent studies have established that exercise stimulates the creation of new brain cells (neurons) in the hippocampus – that part of the brain responsible for learning and memory formation (Gage 1999). Additionally physical movement pumps up existing ones, improves mood, blunts ageing-related memory loss, sharpens decision making, dulls stress, deepens the quality of sleep and if you happen to be a student of any age, improves your grades (Reynolds 2014).

The encouraging news is that we do not have to set our sights on running a mini-marathon, or even joining a gym. As we will discuss, studies on the brain indicate that vigorous exercise is not always necessary for effective functioning. Even a small amount of physical activity can make an enormous difference to the functioning of the brain. Indeed routine everyday activities can provide sufficient opportunity for healthy movement.

If there is one overriding message in this chapter, it is that the challenge of physical movement is to train our brains to make physical activity a habit and thus enhance learning. In this context we will explain why exercise is more about serving and nurturing our brain's learning needs than it is about shrinking waistlines or toning muscles. The English poet, John Dryden (1631–1700), put the essence of this current chapter into seven words: 'The wise, for cure, on exercise depend'.

The connections between any form of physical activity and thinking are intricate, addictive and multidirectional. For thousands of years, philosophers, mystics and scientists have talked about the mind-body connection. The first-century poet Juvenal, praised *'mens sana in corpore sano'* ('a sound mind in a sound body'). But it was not until the past two decades or so, with the advent of brain scanning technologies, that we zeroed in on the activity of individual brain cells that scientists began to understand just how a healthy body makes for a healthier brain at a molecular level.

The capacity to move is a gift; the ability to move is a skill; the willingness to move is a choice. Since humans have existed on Earth, being physically active was essential for survival and this shaped our physiology as we evolved. More recently our ancestors laboured on farms and in factories. But the decline in labour-intensive industries, plus the invention of motorised transport and industrial machinery and other labour-saving devices, means physical activity is less necessary. We are now living our lives in totally different ways, dominated by information technology and devices unimaginable even fifty years ago. In the west, we are the first generation in human history in which the mass of the population must deliberately exercise to remain healthy. In this context, it is crucial to recognise the importance of brain health as a prerequisite to positive learning.

Recent studies have established that the brain, just as much as the body, needs physical activity to develop and function correctly. The negative consequences of a sedentary or inactive lifestyle are becoming clearer. Some scientists even go so far as to say that it is better to be physically fit and overweight than to be of normal weight and sedentary. Indeed, here we may note a cautionary word from Edward Smith-Stanley (1873): 'Those who do not find time for exercise will have to find time for illness'.

The Brain and Body Are Organically One

Our ability to move in harmony is controlled by our motor system. The starting point for any movement is a nerve signal; the nerve signal instigates either an automatic movement, such as I am going to stand up, or a goal-oriented movement, for example, I am going to practice violin. When we make an automatic movement, the planning begins in the cerebral cortex or the thinking part of the brain. The intention to perform the movement is then transmitted from the cortex to the basal ganglia. The basal ganglia helps to select the appropriate motor sequence, and then signals are sent from the basal ganglia to the motor areas of the cortex and the brainstem allowing us to move, and finally the basal ganglia monitors the movement.

While the basal ganglia are involved in selecting and initiating motor responses, the cerebellum is involved in fine-tuning and coordinating movements to ensure accuracy. The cerebellum, whose name means 'mini-brain' or 'little brain', is an outcrop of the brainstem which receives a constant stream of incoming sensory information that enables it to monitor the position of various parts of the body. Receiving inputs from the eyes and ears, it despatches instructions back through the brain stem to other regions in the brain. This part of the brain is like the conductor in an orchestra. It stores learned sequences of movement and coordinates and fine-tunes messages from elsewhere to create fluid body movements. It is where we place most of our habits which we perform daily without too much thought - like driving, walking, typing and so forth.

Notice how babies learns to walk. They start by learning each movement separately and slowly, relying on feedback from their muscles and joints to tell them whether they are getting it right or wrong. This is inevitably a slow process at first, because it takes time for nerve muscles to travel between the limbs and the brain. But with practice the child becomes steadier, faster and makes fewer mistakes (Biddle et al. 2015). This happens because the brain does not need such detailed feedback, as it has identified the correct muscles and actions required for walking. We acquire any skill like this by a slow and deliberate trial and error process.

Neurogenesis: Building a Better Brain

As we have seen, neurons communicate with each other via their axons. The most important feature of our brain is the multiplicity and complexity of the connections between the neurons. The human brain has billions of neurons and trillions of connections. This allows us to move, think fast, and do three essential mental functions associated with being uniquely human: learning, using language and thinking in an abstract way.

As mentioned in Chapter 10 our brain needs a constant supply of glucose-derived sugar. We must obtain this sugar from blood pumped by the heart to the brain. The brain depends entirely on blood to give it the energy it needs to produce chemicals that are critical for learning and memory. The largest proportion of energy in the brain is consumed by our neurons when they are busy processing incoming sensory information or helping us think through difficult problems.

For centuries, people believed that the brain was a relatively rigid, inflexible organ that did not change much structurally over a person's life span. It supposedly had no ability to make new neurons. In school biology classes, most of us were taught that we were born with a certain number of brain cells and would have only those and no others for the rest of our lives. The assumption was that adult brains possessed all the neurons that anyone would ever need.

All this changed in the 1990s, when neuroscientists at the Salt Lake Institute for Biological Studies made a remarkable discovery. Mice raised in an 'enriched environment', in which they had running wheels, toys and ladders, with lots of opportunity for physical activity, grew new fresh neurons, compared to mice housed in standard laboratory cages, with no such opportunities for active movement. The research team under the direction of Professor Fred Gage was surprised to find that the mice reared in an enriched environment showed evidence of making new cells in their brain.

Before being euthanised, the mice had been injected with a chemical compound that incorporates itself into actively dividing cells. During autopsy, these cells could be identified by using a special dye. Gage and his team presumed they would not find such cells in the mice's brain tissue, but to their astonishment they did. Up until the point of death, the mice had been creating fresh neurons. Their brains were regenerating themselves.

All the mice showed this vivid proof of what is known as neurogenesis or the birth of new neurons. But the brains of the athletic mice showed much more. These mice, the ones that had scampered on running wheels, were producing two to three times as many new neurons as the mice that had not exercised.

But does neurogenesis also happen in the human brain? To find out, Professor Gage and his colleagues obtained brain tissue from deceased cancer patients who had donated their bodies to research. While still living these people had been injected with the same kind of compound used on Dr Gage's mice (pathologists were hoping to learn more about how quickly the patients' tumour cells were growing). When Gage dyed their brain samples, he again saw new neurons. Like the mice, humans showed evidence of neurogenesis, and this neurogenesis was centred almost exclusively in the hippocampus.

Since then, scientists have been finding more evidence that the human brain is not only capable of renewing itself, but that physical exercise speeds up the process. Another group of neuroscientists at Columbia University set out to determine if something similar was happening in living humans. They gathered a group of men and women ranging in age from 24 to 45 and asked them to begin working out for 1 hour 4 times a week. After 12 weeks the participants in the study, not surprisingly showed improvements in their fitness level, but more significantly they found evidence that the human brain is capable of renewing itself by giving birth to new neurons. All movement is helpful, but aerobic or vigorous exercise speeds up the process.

But something else happened because of all those workouts: blood flowed at a much higher volume to the hippocampus part of the brain where neurogenesis occurs. Using a Functional fMRI machine (which measures the size and shape of the brain) showed that a portion of each person's hippocampus now received almost twice the blood volume it had before. Scientists suspected that the blood pumping into that part of the brain was helping to produce neurons there.

Furthermore the Columbia study suggests that shrinkage of the hippocampus as we age could be slowed via exercise. The volunteers in this study showed significant improvements in their memory, as measured in a word-recall test, after they had been working out for three months. And moreover, those with the biggest increase in fitness levels had the best score on the test of all the participants.

Like the mice, this study suggests that humans gain a constant supply of new neurons in the brain. Exercise it seems stimulates the creation of new neurons. At the end of the study Professor Gage concluded that our behaviour could control and change the structure of our brain. The discovery of neurogenesis has huge implications for the role of exercise, brain health and learning.

Neurogenesis and Exercise

Neurogenesis is the process that develops and maintains the capacity of the brain to function by replacing neurons that are damaged or killed. In this regard maintaining synaptic connections is essential for brain health and learning. There are several factors that contribute to the damage or death of neurons, including all forms of cardiovascular disease, diabetes, drug abuse, anxiety and most especially stress and alcohol.

The process of neurogenesis is regulated by a hormone called brain-derived neurotropic factor (BDNF). A rather strange name but think of brain-derived neurotropic as a protein – some neuroscientists refer to it as 'brain fertilizer' or as 'miracle grow' for the brain. As a regulatory hormone it plays a key role in making new neurons, as well as improving the connections between existing neurons. BDNF gives the synapses the tools they need to take in information, process it, remember it and put it in context (Ferguson 2017). Exercise also increases an important chemical vascular endothelial growth factor that supports the growth of blood vessels in organs and tissues throughout our body and brain. More blood vessels mean more blood flow and more blood flow means more oxygen and nutrients sent to our brain. In other words exercise prepares neurons to connect, while mental stimulation allows our brain to capitalise on that readiness.

Neurogenesis is a process, and glial cells are critical to how it works. For the first seven days of their life, young neurons will soak up nutrients from nearby mature neurons. When these neurons are excited (through regular exercise), the new neurons get excited at the same time. Physical activity helps to make new neurons proliferate. As a result the new neuron encodes information from their adjacent mature neurons – in other words, they form connections with the help of glial cells and myelination, and in turn they become mature neurons themselves after a twenty-eight-day habit formation process. To read more about the process of BDNF take a short detour into the inner working of neurons and read the case study on glial cells below.

Case Study: Glial Cells

Our brain contains about 100 billion or so neurons. It also contains trillions of glial cells, ten times the number of neurons. These take their name from the Greek word 'glue' because they appear to stick to neurons. They have a star-like shape. For a long time, glial cells were thought of as the poor relations to neurons, and some scientists referred to them as Cinderella cells because they are preoccupied with housekeeping tasks. Some glial cells play an important role in repairing damaged neurons. They can move about the brain, homing in on damaged tissue and cleaning out debris from injured cells. They also maintain the cerebrospinal fluid which bathes the brain like scaffolding to prevent neurons from slithering around. The most common glial cells cluster round the neurons, like dutiful attendants. They play a crucial role in forming a protective sheath that insulates the axons that carry electrical signals.

Neuroscience research now indicates that regular movement has a demonstrable benefit. In the past decade, scientists have made huge strides in deciphering the extraordinary relationship between physical activity and overall health. Adopting an 'active lifestyle' has been recognised as a brain enhancing activity. Staying 'brain healthy' does not require a huge investment of time and money.

In the US, Project Active aimed to establish if an active living approach to promoting 'movement' could be as effective as a traditional fitness-oriented approach in health promotion. To test this hypothesis, the project recruited 116 men and 119 women living in Dallas (average age 46 years). The research involved measuring subtle changes in the body over time. The participants were randomly assigned to one of two groups. One group was labelled 'Structured Exercise' and followed a traditional gym-based approach. They were supervised for six months and then paid a fee to remain a member of the gym for up to 2 years. The second group were labelled 'Active Living' and when they met they did not exercise but had group discussions designed to assist adoption and maintenance of active living, such as goal setting and self-monitoring strategies for self-reward and reinforcement, building social support and relapse prevention techniques. At first, the meetings were weekly (first 4 months); these then tapered to bi-weekly (from 4 to 6 months); monthly (6 to 18 months); and finally, tri-monthly from 18 to 24.

After 6 months, both groups had improved cardiorespiratory fitness, but the structured group had achieved more. In addition, both groups had reduced total cholesterol and percentage body fat. After twenty-four months, both groups had, again made significant improvements in overall health and feelings of well-being. For instance, both groups showed higher changes in oxygen uptake, as well as total energy expenditure (those with the best scores were those with the most consistent activity). The authors drew the conclusion that the active living approach was as effective as a traditional fitness approach in improving overall health. What was interesting is that many of those in the active living group moved to more structured physical activity after a year.

Active living can therefore be defined as taking everyday opportunities to build conscious body movement into daily routines. This can be something as simple as parking your car a little further from your intended destination; walking briskly from the car park to where you work or taking the stairs rather than the escalators. We think this finding is encouraging, as it might inspire people who would be intimidated about the idea of 'exercising' to just get up and move.

Exercise and Neurogenesis

The concept of physical activity can be classified as having three distinct categories: light, aerobic and vigorous activities. This can be represented as a physical activity pyramid. What makes each of the three levels in the pyramid different is the intensity of the activity in terms of movement and the amount of physical energy required to perform the different tasks. Intensity refers to the amount of oxygen consumed and the subsequent boost of blood flow to the brain. An easy way to estimate intensity is through a method called the 'talk-sweat-test'. Basically the less you can talk due to shortness of breath the more oxygen will circulate in your brain; the more you sweat the greater the number of neurotransmitters is released in your brain.

Light Physical Activity

The two words most associated with light physical activity, as represented in the first level in the pyramid, are 'gentle' and 'consistent'. Leisurely walking is a good way to support a healthy lifestyle and encourage optimum brain function. Neuroscientists have recently found that the foot's impact during walking sends pressure waves through the archeries that significantly modify and can increase the supply of blood to the brain. Walking is thus an advanced brain exercise.

How much physical activity do people need to gain health benefits? Originally standards were set concerning cardiovascular fitness in healthy adults but were later amended in the light of new research evidence which indicated that the volume of activity was more important than the frequency. The World Health Organisation (WHO) recommend that all adults should achieve a minimum of 150 minutes of moderate intensity physical activity a week in bouts of at least 10 minutes.

In more recent years, guidelines have been extended to include the connection between brain health and movement as an outcome. Consider the results of an American study carried out in 2005 which followed 25,000 adult men and women to quantify the relationship between lifestyle, learning and heart functioning. They found that 72 per cent of women and 64 per cent of men (aged 40–65) do not participate in any regular physical activity.

As a result of this study, much debate then focused on how *little* exercise is needed to create a beneficial effect. The thinking at the time was that if the goal was lower, the more likely that a largely sedentary adult population might be inspired to work towards it. According to neuroscientists, walking is now considered one of the most beneficial activities for brain health. It is a simple yet powerful activity. It is the most natural physical activity especially for beginners. It is free, easy to perform and does not require any training or equipment.

Elderly sedentary people who began a walking programme showed significant growth in several areas of the brain after six months. Scientists believe that the workout prompted the creation of new neurons, as well as new blood vessels and connections between the neurons. The walkers' brains were bigger, faster and younger, and they consequently performed better on tests of memory and decision making than people who had remained sedentary.

Aerobic Physical Activity

The philosophy of aerobic fitness is simple; if our muscles get slack and unused, eventually our brain will start to copy them. This philosophy also encourages people to think about their brain health now and not just in their senior years. One of the fundamental things we can do to maintain a learning brain is regular aerobic activities. The word aerobic means 'with oxygen' as it is any type of activity that increases our heart rate and breathing with a little pressure on the lungs.

Complex physical activity involving coordination, balance and diverse movements, stimulates multiple brain areas simultaneously, reinforcing the connections between them. Our brain reaps instant benefits when we exercise in novel and unexpected ways. This makes 'cross activity' (e.g. walking, swimming and tennis) far more productive than performing precisely the same exercise day after day. With imagination it is possible to build aerobic activity into our daily routine.

As the brain does not have its own supply of 'fuel', it is dependent on the circulation of the blood to supply it with sufficient oxygen and nutrients to help neurons communicate better with each other as well as helping to create new neurons. Aerobic activity pumps more blood to the brain and it helps it use glucose more efficiently. Have you ever experienced a 'feel good' sensation immediately after exercise? If so, you have experienced the phenomenon known as 'runner's high'. It is a temporary euphoric state that typically involves elation, contented feelings and a great sense of well-being. With aerobic activity, the body releases chemicals called endorphins.

Endorphins are often called the body's painkillers and are the natural morphine produced in the brain. With aerobic activities, they flood the body and increase key neurotransmitters like serotonin and dopamine, which support our energy level. More importantly, they help to clear away negative emotions to improve one's mood. They are the chemicals that produce the feeling of well-being and relaxation that also enhance our ability to deal with stress and fatigue, as well as building resilience.

Vigorous Physical Activity

The third level in the pyramid is known as vigorous physical activity. This usually refers to more structured leisure-time physical activity, such as jogging, swimming, 'keep-fit' activities, running and stepping on machines, hiking on a nature trail, weight-lifting and recreational sports.

Strength training like light weightlifting or stretching, strengthens bones, affects stamina and cultivates flexibility. It does not just target specific muscles but the surrounding muscles also. Moving against something causes the muscles to contract with an external resistance leading to an increase in strength, tone, or muscular endurance as in swimming for example. Muscle strength becomes more important the older we get, as we naturally lose muscle mass. Strength peaks between thirty-five to forty-five years and after that we start losing about 1 per cent of our strength each year (due to chronic disuse rather than muscle ageing). The maintenance of muscle mass and strength may decrease.

In 2012 a study performed in Brazil revealed that if we can sit on the floor and then rise to a standing position, it can tell us something about our overall muscle strength and fitness. It turns out that if we can get yourself up from the floor using just one hand – or even better, without the help of any hand, then we are in the top 25 per cent of muscle-skeletal fitness. Put simply, the better we can do this task without relying on our hands for stability and support, the longer we will remain healthy and active.

Since the 1990s, there has been considerable increase in the knowledge about how physical activity may prevent or delay brain degenerative diseases, such as Alzheimer's and Parkinson's. Because Alzheimer's is characterised by a reduction in the number of neurons in brain areas such as the hippocampus and because physical activity supports neurons in this part of the brain, exercise may help to slow the progression of the disease. There is now convincing research evidence from the neuroscientific community that vigorous activity just twice a week halves our risk of general dementia and reduces our risk of Alzheimer's disease by 60 per cent. Research shows that people with Alzheimer's who enrol in a 12-week moderate exercise programme show improvement in brain functioning, particularly memory.

When an activity is sufficiently vigorous it alters the structure of the brain. The volume of the brain usually shrinks as we get older, and less brain means less power to think, reason, and do everything we need to do. But vigorous activity slows this shrinkage. Recently neuroscientists found that the size of the hippocampus of older adults who walked for 40 minutes, 3 times a week, for 1 year increased by 2 per cent. Even later in life exercise can protect and improve the structure of the brain.

Resources

The Professional Development Service for Teachers (PDST) produced a resource called *Movement in the Classroom: Movement Breaks and Energisers* for teachers in 2020. The resource outlines short movement breaks and activities that can be used during the school day. The resource is available here: <https://www.pdst.ie/sites/default/files/Movement%20in%20the%20Classroom%2020.10.20.pdf>

Movement and Stimulation of Learning

We know that movement underpins the development of fine motor skills, physical fitness, language and communication, self-esteem, confidence and learning (Macvier et al. 2019). Furthermore engaging in physical activity increases blood flow to the brain, ensuring a steady supply of oxygen and nutrients. This improved circulation supports overall brain function, including cognition and memory. Exercise stimulates the production of new neurons in the hippocampus, the centre for learning and memory.

In 2010 a group of scholars in the US conducted research with 4th grade children, using control groups to compare those who were permitted a 15 -20 - minute recess period during the day, with those who did not have any recess. They found that sixty percent of children benefited considerably. They displayed more 'on-task' behavior or fidgeted less (or both) on recess days (Jarret et al. 2010). Movement breaks also provide important opportunities for learners to collaborate, cooperate and socialize with each other so that they build social and emotional connections in the classroom.

The *National Council for Special Education* in Ireland (2020) has advocated for the incorporation of movement breaks into the classrooms, to support students to maintain their concentration and attention levels (Hoza et al. 2015) and to regulate their energy (Mac Cobb et al. 2014). Movement breaks are short breaks of 10 -20 minutes allocated to physical activity in the classroom. During these breaks, the brain is able to process the new information learned before engaging in the next lesson. Movement breaks provide a viable approach to improve overall fitness, cognitive function, and ultimately academic achievement (Donnelly and Lambourne 2011) by encouraging a positive attitude towards physical activity and movement within the classroom context. And interestingly physical activity amongst children leads to an immediate boost in concentration (Mahar et al. 2006; Konigs et al. 2014; Caternio and Polak 1999).

Conclusion

There is sufficient evidence to indicate that exercise and physical activity have considerable health promoting benefits for an individual. Physical activity increases health for many different organ systems and helps to limit risk of disease across the lifespan. We have established that physical movement and participating in regular physical activity is exceptionally good for the ageing brain. If physical exercise is an option, choose open-skill activities that keep us on our toes (such as tennis, football or badminton) and require us to constantly adapt to a changing environment. In contrast closed-skill activities such as running or swimming which are very predictable and totally within our control. The former lead to a bigger BDNF boost which will promote the survival and growth of neurons.

Such is the power of exercise on the brain that some scientists believe we might be thinking about it back to front. As one health expert writes 'I believe we should stop talking about 'exercise' altogether and start thinking instead about 'movement'. We simply need to move more during the day, throughout the day. We need to design our lives around movement. We are designed to be active, but modern life makes us sit for hours at a desk, in cars, in front of the TV' (Chatterjee 2018: 152). Dan Buettner describes the daily habits of the world's oldest living people. In terms of exercise, there is one big takeaway: such people don't necessarily do intense sweat sessions. They simply build low-intensity exercise into their everyday lives. This research along with many other findings clearly shows that having a fit body can lead to a healthy brain, as exercise can give us all sorts of cognitive boosts.

References

Biddle, S., Mutrie, N., Gorely, T. and Faulkner, G. (2015). *Psychology of Physical Activity: Determinants, Well-being and Interventions* (3rd edn). London: Routledge.

Buettner, D. (2010). *Blue Zones: Lessons for Living Longer from the People Who've Lived the Longest.* London: National Geographic Society.

Caterino, M. C. and Polak, E. D. (1999). Effects of Two Types of Activity on the Performance of Second-, Third-, and Fourth-grade Students on a Test of Concentration. *Perceptive Motor Skills*, 89(1), 245–248.

Chatterjee, R. (2018). *The 4 Pillar Plan: How to Relax, Eat, Move, Sleep Your Way to a Longer, Healthier Life.* London: Penguin Random House

Donnelly, J. and Lambourne, K. (2011). Classroom-Based Physical Activity, Cognition and Academic Achievement. *Preventive Medicine*, Vol 52(1), 36–42.

Dryden, J. (2016). The Secular Masque. In S. Ratcliffe (Ed.), *Oxford Essential Quotations* (4th edn). Oxford: Oxford University Press.

Ferguson, A. (2017). *Fitness: This Is Your Brain on Exercise.* New Press. Available from <https://eu.news-press.com/story/life/wellness/angie-ferguson/2017/04/03/fitness-brain-exercise-angie-ferguson/99834536/>

Gage, F. (1999). Running Enhances Neurogenesis: Learning and Long-Term Potentiation in Mice. *PNAS*, 96(23), 13427–13431.

Hoza, B., Smith, A. L., Shoulberg, E. K., Linnea, K., Dorsch, T. E., Blazo, J. A., Alerding, C. M. and Mccabe, G. P. (2015). A Randomized Trial Examining the Effects of Aerobic Physical Activity on Attention-deficit/Hyperactivity Disorder Symptoms in Young Children. *Journal of Abnormal Child Psychology*, 43, 655–667.

Jarrett, O., Maxwell, M., Dickerson, C., Hoge, P., Davies, G. and Yetley, A. (2010). Impact of Recess on Classroom Behaviour: Group Effects and Individual Differences. *The Journal of Educational Research*, 92(2), 121–126.

Konigs, M., Oosterlaan, J., Scherder, E. and Verbeurgh, L. (2014). Physical Exercise and Executive Functions in Preadolescent Children, Adolescents and Young Adults: A Meta-analysis. *British Journal of Sports Medicine*, 48(12), 973–979.

Livingston, G. et al. (2020). Dementia Prevention, Intervention, and Care: 2020 Report of the Lancet Commission. *The Lancet*, 396 (10248):413-446.

MacCobb, S., Fitzgerald, B. and Lanigan O'Keeffe, C. (2014). The Alert Program for Self Management of Behaviour in Second Level Schools: Results of Phase 1 of a Pilot Study. *Emotional and Behavioural Difficulties*, 19(4), 410–425.

Maciver, D., Rutherford, M., Arakelyan, S., Kramer, J. M., Richmond, J., Todorova, L., et al. (2019). Participation of Children with Disabilities in School: A Realist Systematic Review of Psychosocial and Environmental Factors. *PLoS ONE* 14(1). <https://doi.org/10.1371/journal.pone.0210511>.

Mahar, M., Murphy, S. R., Sheilds, A. and Readeke, T. (2006). Effects of a Classroom Based Program on Physical Activity an On-task Behaviour. *Medicine and Science in Sports and Exercise*, 38(12), 2086–2094.

Nerurkar, A. (2024). *The 5 Resets: Rewire Your Brain and Body for Less Stress and More Resilience*. Dublin: Harper Collins Publishers.

Rahey, J. (2008). *Spark: The Revolutionary New Science of Exercise and the Brain*. New York: Little, Brown.

Reynolds, G. (2014). *The First 20 Minutes: The Surprising Science of How We Can Exercise Better, Train Smarter and Live Longer*. London: Icon Books.

Smith-Stanley, E. (1873). *The Conduct of Life*, an Address at Liverpool College, UK. 20 December 1873.

Strauch, B. (2011). *The Secret Life of the Grown-Up Brain: Discover the Powerful Talents of the Middle-aged Mind*. London: Penguin Group.

CHAPTER 12

Sleep and Brain Health

> 'Your life is a reflection of how you sleep, and how you sleep is a reflection of your life'.
>
> Edward Smith Stanley (1973)

Sleep is a dynamic process that supports various cognitive functions, including learning and memory. Having sufficient quality sleep is essential for the ability to learn and retain new information. Whether we are under pressure trying to learn a new subject topic or studying for a big examination, depriving ourselves is a counterproductive strategy. While we still do not know exactly why we sleep, what is clear is that the sleeping brain is far from a resting brain.

The science of sleep has exploded in recent years, revolutionising our understanding about previous notions that our bodies and brains are inactive during sleep. In some ways during sleep the brain works even harder. Good sleep is fundamental to just about everything to do with brain health – clear thinking, strength, energy, ability to relate to others, as well as being one of the best means to manage stress and anxiety. Sleep is essential for every cell in the body and critical for the nervous system, especially the brain. This is because it is the nervous system which controls and co-ordinates all the other systems of the body, such as the immune system, the digestive system, the glandular system and reproductive system. In short sleep recharges the physical batteries and rejuvenates the brain.

In this chapter we will try to unravel some questions about what goes on in our brain while we sleep and why it is so essential to our sense of well-being, particularly for our mental health. What are the barriers to sleep? How much sleep is 'normal' to function effectively? Why do we sleep, and is there a best time to sleep? How does sleep consolidate learning? What practical things can one do to sleep deeply most nights? How does sleep research directly benefit our lives? What follows will try to answer some of these questions.

The Importance of Sleep for Performance

The electric lightbulb radically altered the world. Thomas Edison's invention has allowed billions of people to spend 24 hours of the day as they please at night while their ancestors had no such freedom. We now live in a world where we are completely dependent on electric power to go about our daily routines. Despite all the obvious benefits, there is one major drawback: artificial light has fractured sleep.

We have been falsely led to believe that the importance of sleep has been undervalued. Thanks to the knowledge gleaned over recent of the undervaluing of significant of sleep has been totally revised. Why we spend so much time needing sleep is still largely a mystery. In certain quarters it is common to hear some people publicly boast about only getting 4 or 5 hours a night and that sleep is the 'enemy'. However, scientists have made fascinating discoveries about what exactly is going on while we sleep that supports brain health.

'Imagine you could do one simple, pleasant thing that would protect you from heart disease, flu, infections, dementia, mental ill health and car crashes. It would make you more productive, energetic, creative and socially adept. It would make you lose weight and look younger' (Walker 2004: 705). Thus, for Walker all these things become possible. Sleep is necessary if we want to be able to learn. Sleep has been shown to be crucial for memory consolidation. During certain stages of sleep, the brain engages in processes that enhance the consolidation of memories. This is why getting adequate and quality sleep is important for effective learning. As Arianna Huffington (2016) points out, if we value our brains, then we need to get more sleep.

When we think of our fundamental human needs and urges we usually do not think about sleep? But the need to sleep is just as critical to our wellbeing as food, water, security, love and sex. In fact, the drive to sleep is so strong it will supersede the drive to eat. Our brain will just go to sleep eventually despite all our conscious efforts to keep it at bay. As sleep researcher Allan Rechtschaffen (1971) once observed, 'if sleep doesn't serve an absolutely vital function, it is the biggest mistake evolution ever made' (cited in Mignot 2008: 106). Is there anything that feels more welcome, than crawling into bed when we are exhausted?

We spend a third of our lives asleep and a good night's sleep is integral to our health and well-being. The amount we need varies throughout the life-span. While scientists recommend between 7 and 9 hours of sleep each night, this is a rarity for many adults today. Although there is no agreed definition of sleep, neuroscientists generally agree that sleep is something that happens when conscious activity modulates into cycles of different rhythms. They also tell us that getting to sleep is an acquired habit, sometimes easily lost and hard to regain.

Sleep helps prime our brain to be receptive and alert, to be ready to learn and acquire new information. After learning, sleep helps to cement those facts and skills and stops us forgetting them. Memories move from deep within the brain's hippocampus to the higher-level cortex layers – and this transportation happens overnight during sleep. Sleep scientist Matthew Walker discovered that it is the lighter NREM 2 sleep that is responsible for memory refreshment and the deep NREM sleep that provides the best memory retention. Before we get on to how we use sleep to consolidate learning we need to understand why we sleep and what happens to the brain in a normal night's sleep.

Why Do We Sleep?

The reasons for sleep remain one of the most profound mysteries. If the function of sleep is purely physiological, what exactly does it do? If it is psychological, what is its purpose? If it is neurological, in what way does the brain benefit? These questions and others have intrigued philosophers and scientists for centuries (Hobson 2005).

The first obvious idea is that we sleep to be renewed, refreshed and energised. Our sleep brings rest, renewal and energy, not only to our bodies, but also to our hearts, minds and spirits. Sleep relieves us from the stresses of daily life. It builds us up, making us stronger, more vibrant, more productive and alert individuals. This view of sleep is known as the restorative theory, in which sleep is seen as a way of restoring the resources we use during the day.

In earlier times a major function attributed to sleep was that of healing. In psychoanalysis for instance, Freud claimed that dreams 'are the royal road to the unconscious' – a process of accessing the non-conscious. By this he meant the process of accessing the non-conscious parts of the mind. He argued that repressed fears and desires, and other contents of the unconscious are allowed to emerge and play themselves out during dream sleep. This process is important for self-awareness, as it allows conflicts and anxieties to be explored and can contribute to personal healing. This may explain why most languages have an expression that means something like 'sleep on it'.

Closely associated with the healing idea of sleep is the role of innovation and creativity. Our brain is buzzing all the time with electrical activity creating and producing brain waves. Creative ideas are most likely to happen not when our brain is hard at work but when it is relaxed. A relaxed state during sleep opens up the mind to new possibilities, allowing unexpected thoughts and original ideas come to the surface and allowing for a time when we solve puzzles and discover new insights. As frequently reported by scientists, inventors, artists, musicians and writers when talking about some of their more ingenious creations, a particular dream has often sparked an unusual connection in seeing something new or in a different way. As Ray Charles, the legendary pioneer of soul music said, 'Dreams, if they're any good, are always a little bit crazy' (Charles 2004).

Another idea suggests that a major function of sleep is to calm the mind by cementing learning and memories and to prune unnecessary neural connections. This idea is known as the housekeeping theory. During the day, the brain is bombarded and consequently overloaded with stimulation. During sleep deep cross-connections are formed in the brain. The idea is that sleep (especially during episodes of REM sleep) evolved so that connections between neurons in the brain can be whittled down overnight, making room for fresh memories to form the next day, thus allowing the brain to 'let off steam'. The majority of neurons weaken during slumber - this process may have the advantage of consolidating learning and memory. If this theory is right, it would explain why when we miss a night's sleep, we find it harder the next day to concentrate and learn new information, since we may have less capacity to encode new experiences.

Others suggest that sleep's primary purpose is the consolidation of learning. Studies suggest that the brain uses a night's sleep to consolidate the actions and skills learned during the day. These studies suggest that sleep plays a critical role in storing important physical and intellectual experiences as well as making subtle connections that were invisible during waking hours. Just think about the insane volume of neural connections the brain has to make in the course of any given day. At some point we have to decide which of these connections are worth holding on to, and which can be ignored. When we turn off the lights, that's when the brain begins to sort out the meaningful from the trivial.

The most recent, and some say most exciting theory, is called the maintenance and repair theory. Sleep is not just about sleep, but rather while we sleep the brain has a chance to catch up, to clean out and recycle the toxins that have built up during the day. A remarkable discovery in 2013 by Maiken Nedergaard found that the brain has its own self-cleaning up drainage system called the glymphatic system, similar to the lymph vessels that drain waste products from tissues in the rest of the body. During deep sleep the brain swells, as intra-neuronal spaces expand. Toxic proteins are washed out of the brain into the cerebrospinal fluid (CSF) during the night. Arianna Huffington describes this as bringing in the overnight cleaning crew to clean the toxic waste proteins that accumulate between neurons during the day. Failure to get enough sleep and clear out toxins is linked to a host of degenerative diseases (Nedergaard 2013). This theory thus holds that sleep is a function of the development and ongoing maintenance of the brain itself.

These theories of sleep teach us something unique about the workings of the human brain. We have all seen models in which the brain is presented as a map with clear territories: this region does this while that region does that. But such a model ignores the most important aspect of the human brain, that no brain region works alone. Instead like a city, areas work in a constant clamour of crosstalk and negotiation and cooperation. Even while particular skills can be restricted to local brain regions, creativity and learning are whole-brain experiences, arising from the sweeping collaboration of distant neural networks while we sleep.

The Science of Sleep

It was not until the late 1920s, with the development of human electroencephalography (EEG), that scientists could study the brain's electrical activity during sleep. EEG recordings, often called *brain waves*, revealed that sleep is not a single, one-dimensional state, but rather a dynamic process during which the brain continues to respond to the environment and the internal functions of the body, while still overseeing the operation of sleep itself.

In the 1950s the significance of rapid eye movement (REM) during sleep was discovered and unique stages of sleep were fully described (Aserinksy 1996). Sleep scientists now say that sleep is an active complex state and has quite distinct characteristics based on eye movement known as non-rapid eye movement (NREM) and rapid eye movement (REM). (Aserinsky and Kleitman 1953). A normal night's sleep consists of a recurring pattern of sleep cycles. Each sleep cycle contains stages that last for about 90–110 minutes on average. Throughout the night an individual may go through several complete sleep cycles. The quality and quantity of sleep cycles are essential for maintaining brain health.

Stages of Sleep

Sleep is a distinct state in which consciousness is suspended and a person is mostly, but not completely, unresponsive to the outside world. It is different from resting, because during sleep the muscles are relaxed, heart rate is reduced and activity in the brain shows characteristic changes.

Stage 1: can be viewed as a transition between wakefulness and sleep and characterised by drowsiness. A typical sleeper will spend only about 5 minutes in Stage 1. Our muscles and eye movements are normal as we make ourselves comfortable for sleep.

Stage 2: is known as 'light sleep' and lasts for about 20 minutes. An EEG would show short bursts of brain-wave activity as we enter into a slightly deeper sleep characterised by the slowing of our heartbeat and the lowering of our core body temperature. At this stage it is easy to be dragged back to wakefulness.

Stage 3: involves brain waves that are much slower (this is also known as slow-wave sleep). For this reason it is called 'quiet sleep' as eye and muscle movements have nearly ceased and one would be very difficult to wake. The depths of this third stage contain some of sleep's most fascinating mysteries and oddities, like sleepwalking and sleep talking.

Stage 4: is our deepest phase of sleep, where our bodies recover from the daily grind and muscle movements nearly cease ('we are dead to the world'). The EEG monitor would show our brain producing delta waves – the slowest-frequency brain waves. It is in this restorative stage of 'deep sleep' where human growth hormone is released, the brain detoxifies, and memory is consolidated. If awakened during this deep sleep, we would be confused momentarily and feel groggy for a while. Ideally we need to spend about 20 per cent of our time in Stage 3 and 4 during the night.

Stage 5: is REM sleep, or active sleep, characterised by rapid eye movement. REM sleep starts about an hour and a half after we fall asleep. This stage is usually seen as distinct from the other four stages of sleep, because our brain is highly active and the brain waves displayed on an EEG reading become faster in frequency, resembling those shown during wakefulness. It is in REM sleep that we do most of our dreaming. If we wake up during this phase, we are more likely to remember our dreams.

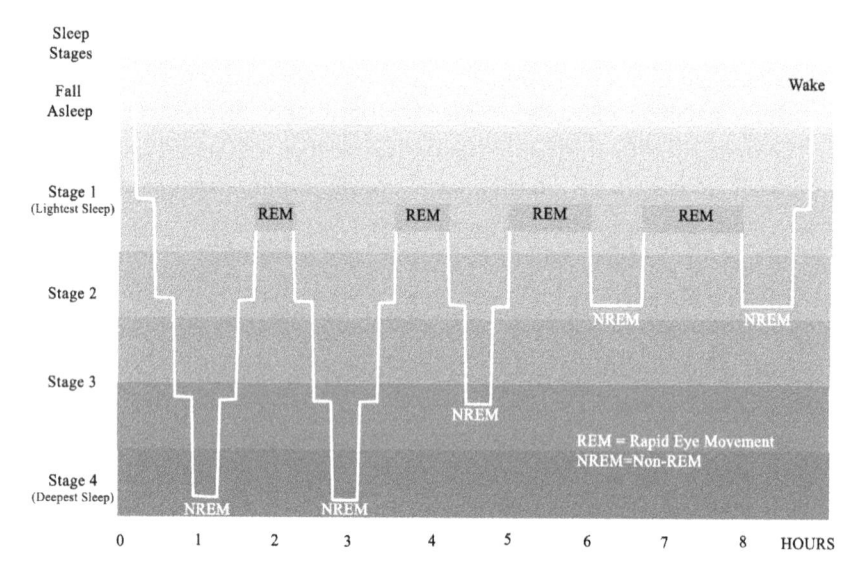

Figure 16: Stages of Sleep – Image by *Dmitry*/<stock.adobe.com>

Adults tend to spend 75 per cent of their time in non-REM sleep and 25 per cent of their time in REM sleep. The work of Censor et al. (2006) at the Weismann Institute in Israel has shown that REM sleep improves the learning of a task. They measured the length of time it took 4 people to learn how to recognise a pattern of stripes. The people were tested both before and after REM sleep. If they were woken before or after REM sleep, they failed to master the task. In other experiments conducted with rats, they found that interrupting REM sleep completely thwarted the learning of a new task, yet interrupting non-REM sleep just as frequently did not. So it appears a good night's sleep acts as a kind of cement for our learning.

The night's sleep described above is the ideal, but there are times when sleep is elusive, when we have difficulty falling asleep, experiencing nights of tossing and turning, drifting in and out of consciousness or having some other sleep-related problems. This makes sense because sleep is a complex process. As a general rule when it comes to sleep there is no such thing as normal. Each night is different, as how much sleep we need or get varies from person to person.

Sleep and Brain Health 205

How Is Sleep Regulated?

Deep inside the human brain is a living clock that acts as our body's timekeeper, telling us when to wake, sleep, rest and play. We each have an internal body clock, corresponding to one day, dividing day and night, called our circadian rhythm. Our circadian rhythm influences almost every aspect of our lives, from mood, emotions and core body temperature to metabolism, to when we eat and drink, and the release of hormones. In the 1970s scientists discovered that in humans our circadian rhythm is governed by a small cluster of neurons alongside the hypothalamus, just above the optic nerves in the brain, to help us sense light and darkness. Another way to think of it is as a master clock set to a schedule of approximately 24 hours, the exact time varying for each of us (Littlehales 2016; Russell 2022).

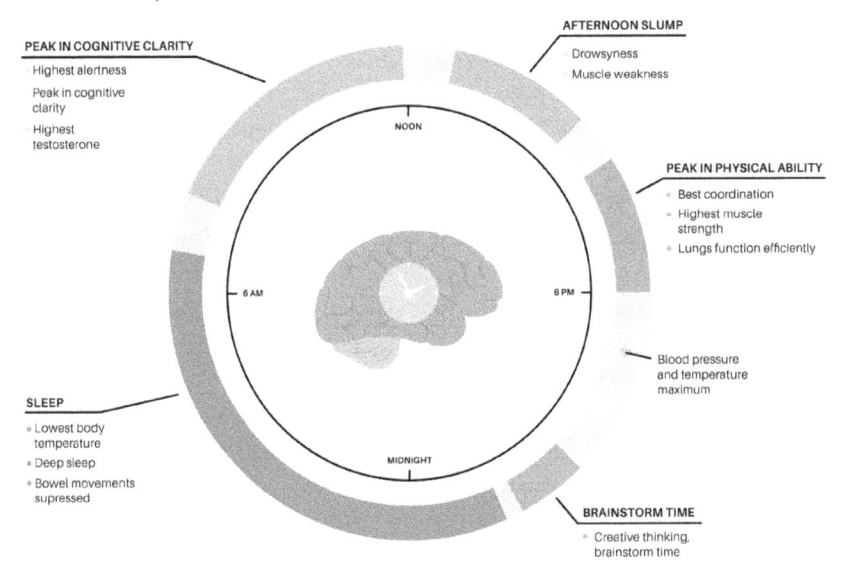

Figure 17: Circadian Rhythm – Image by *pikovit*/<stock.adobe.com>

It is vital to understand that these rhythms are ingrained within us – they are part of the fabric of each and every one of us. They are the product of millions of years of evolution. We cannot unlearn these rhythms as according to Hauri et al. (1996: 147) even if we moved underground and lived without daylight, they would persist within us. Normally the rise and fall of the sun affects how the natural cycles are activated. A typical circadian rhythm, which describes what our body wants to do naturally at various points throughout the day, looks something like the pattern outlined in Figure 17.

Our day-time clock usually begins when light stimulates the eye – a signal is sent to the brain through complex pathways that causes the brain to release cortisol. The cortisol hormone gets us going and prepares the body's temperature for movement and other daily activities. As the sun continues to move across the sky, cortisol levels continue to rise, peaking around 9 a.m. or 10 a.m. At this time a person is most alert (but this may vary considerably (depending on the time of the year and culture).

Cortisol levels reach their maximum around 12 p.m. when energy vision, hearing, smell and taste are at their peak. Cortisol begins to drop in the afternoon (the mid-afternoon slump) and continues to gradually drop further till about 5 p.m. It rises again between 5 p.m. and 9 p.m., during the early evening 'transition phase' when people tend to engage with family and leisure activities.

Once the sun goes down we can see that melatonin secretion starts. Melatonin, a hormone that regulates our sleep, is produced in the pineal gland which responds to light. Once it has been dark for long enough, we produce the melatonin to ready us for sleep. In the evening our bodies are designed to wind down, and by 10 p.m. a feeling of tiredness generally begins and the body begins its repair cycle.

If we keep 'regular' hours and get up in the morning, our need to sleep peaks at night, which coincides with our circadian urge, producing the ideal sleep window. As we will see when we look at the stages of sleep, during the night, after a day's work, we tend to reach our most effective sleeping period for physical repair and growth around 2 a.m. to 3 a.m. while the subsequent period after this until 6 a.m. is important for the psychological repair cycle.

Many of us only really become conscious of our circadian rhythms if we fly across many time zones and experience jet-lag, which is when our rhythms are out of sync with the local dark-light cycle. Similarly we can easily sense our own circadian rhythm by staying up all night. By 2 a.m. or 3 a.m., we feel extremely tired, and by 5 a.m. or 6 a.m. we may be exhausted. But by 7 a.m. or 8 a.m. we will begin to feel less tired. This resurgence occurs because we are reaching a point in our circadian rhythm where our alertness is beginning to pick up. Interestingly after missing a night's sleep, we will not need a full 8 hours of extra sleep to make up for it as an additional 1 or 2 hours usually will be enough.

Lark, Owl or Hummingbird?

Being aware of our body clocks will allow us to begin to understand why we might be feeling lethargic at certain times of the day or why we might be struggling to get off to sleep. Understanding our circadian rhythm also plays a pivotal role in our ability to learn and study productively. We need to determine whether our biological clock runs at a slightly slower or slightly faster pace than the cycle of the sun in a 24 hour day.

According to American sleep researchers, Smolensky and Lamberg, people fall into three groups: 10 per cent of people are larks, and at their best in the early morning; 20 per cent are owls, at their best late at night; and 70 per cent are hummingbirds, who are comfortable being more flexible. Hummingbirds are best at coping with the odd late night or early start. If you are an owl or a lark, it is particularly important to plan a schedule that ensures you get enough rest (Medina 2008: 157). For example, if you are an owl and you decide to go to bed an hour earlier than usual to try to get more sleep, you may well find you spend that hour lying awake, worrying that you cannot sleep. Ideally it would be more productive to allow your biological clock to dictate the amount of time that you spend asleep, while you take steps to improve the *quality* of that sleep.

Whatever our lifestyle and sleep preference, our brain craves consistency. To be able to fall asleep easily we have to rely upon our brain to switch off our wakefulness controls and at the same time to activate our sleep centres. In people who are not stressed, who have not been overtly active before they go to bed and who are without sleep problems, this process is automatic. One of the founders of *Deep Mind*, Denis Hassabis, says 'We think of sleep as an inefficient use of time, but in fact it is the most efficient use of time in terms of learning and memory' (cited in Huffington 2016: 109).

In contrast, sleep deprivation has been identified with poor exam results, social anxiety, stroke risk, drug and vaccine efficacy, mental health and obesity. According to Foster (2022), shift-work can play havoc with the circadian rhythm. For example, working night's limits exposure to the sun which 'entrains' our circadian rhythms, thus ensuring that 97 per cent of night workers never adapt to a regular sleep routine. The consequences are shocking. Men working nights are up to 6 times more likely to separate in the first 5 years of marriage, while night-shift workers have significantly higher rates of many kinds of neurodegenerative diseases such as Alzheimer's disease and Parkinson's disease.

The US Centres for Disease Control and Prevention estimate that 35 per cent of US adults are getting less than 7 hours a night, while the *Sleep Survey* in the UK found that the average was 6.8 hours (Leader 2019). It is a similar story around the world. Insufficient sleep over an extended length of time is also linked with increased risk of serious stress-related illness such as high blood pressure, obesity, stroke and even heart attack, along with neurodegenerative diseases. Sleep loss has been found to contribute to thousands of deaths in workplace settings such as the Bhopal, Exxon Valdez and Chernobyl disasters, as well as mention related accidents in farms, construction and industrial production. As a result prescription of sleeping pills has been increasing dramatically over the past few decades. Sleep clinics, which were once a rarity, are now a feature of most major hospitals.

Loss of sleep reduces the brain's ability to concentrate, coordinate muscles, make decisions and control mood. Furthermore lack of quality sleep may make us irritable, short-tempered, forgetful, depressed, even paranoid or psychotic. The less we sleep the faster our brains age. In adults the brain ventricles – the chambers that hold cerebrospinal fluid – widen as the brain shrinks, and the grooves and fords of the brain become more pronounced, creating gaps. Research has found the lack of sleep increases the pace of brain ventricle enlargement and decreases cognitive performance (Hauri 1996).

How Much Sleep Is Enough?

Scientists tend to agree that the average adult needs between 7 and 9 hours sleep a night, some needing a little more, some a little less. Children and teenagers need more sleep since they need extra energy for growth. People over 60 years old need less sleep. This might explain why 8 hours has become a rule of thumb. Eight is certainly a magical number. It divides nicely into 24 hours and has been linked to sleep for centuries. In monastic times the sun-dial clock partitioned the day into a third for reading, writing and prayer, a third for physical work to maintain the community and a third for the refreshment of the body (Willett and Barnett 2017).

While 8 hours is an average amount of sleep most people aspire to get per night it is not for everyone. The truth is that each of us is different, and as with most things in life, one size does not fit all. The most alert individuals are those who sleep until they are simply no longer tired. On the other hand, some individuals feel that they have had a full night's sleep after significantly less than 8 hours. Some sleep professionals advocate the idea of viewing sleep in a phased way – a method of seeing sleep, not in terms of how many hours a night we are sleeping, but how many 90-minute cycles per week we are fitting in.

Insufficient sleep can have a significant impact on our ability to concentrate and maintain focus. Sleep deprivation can lead to a decline in attention and vigilance. Individuals may find it challenging to sustain focus on a task, leading to increased distraction and difficulty staying engaged in activities that require concentration. It is also interesting to know that neurotransmitters are mostly manufactured at night while we sleep, and consumed next day when we are awake. Sleep deprivation may limit the number of neurotransmitters the brain can replenish. Even broken sleep can profoundly alter moods, which in turn hinders the capacity of working memory and learning.

Working memory, which is crucial for holding and manipulating information temporarily, is negatively affected by sleep deprivation. This can result in difficulties in processing and organising information, impacting complex cognitive tasks. Sleep plays a role in problem-solving and creative thinking. In order to combat the effects of sleep deprivation on one's physiological health, napping for a short period of time and early in the day can be used to maintain optimal brain function (Guadiana and Okashima 2021).

Conclusion

Getting enough quality sleep is essential for learning. Scientists tend to agree that the average adult needs between 7 and 9 hours sleep a night. It is during this sleep that our brain consolidates learning, encoding the memories of the day by integrating them with existing knowledge. This consolidation process is crucial for long-term retention and understanding.

When a normal pattern of sleeping is either restricted or disrupted our cognitive performance worsens. This can have an impact on our vigilance, sustained attention and working memory. Lots of people restrict their sleep during the working week, choosing to work long hours, or socialise instead of prioritising sleep, but then make up for it at weekends. However the effects of sleep deprivation are cumulative, meaning that chronic sleep deficiency can lead to more severe cognitive impairments over time.

So what are the best ways to ensure a decent night's sleep? Sleep scientists suggest a 'transition time' routine of a minimum of 20 minutes without any blue light can enhance quality sleep. Laptops, tablets and especially mobile phones on the other hand generate lots of short wavelength blue light, which interferes with melatonin production by tricking the brain into thinking it is still light. Drinking alcohol or coffee in the evening makes it harder to get to sleep and to stay asleep and also has a knock-on effect on melatonin production the next day, perhaps setting us up for a second bad night. This might be because of the way alcohol is metabolised, releasing chemicals that act as stimulants when our body and brain should be resting.

It appears that our subjective sense of age is influenced by the quality of our sleep and if we feel younger we are better able to have associated benefits like taking on new learning experiences, being socially active and physically active.

References

Aserinsky, E. (1996). The Discovery of REM Sleep. *Journal of the History of Neuroscience*, 5, 213–227.

Aserinsky, E. and Kleitman, N. (1953). Regularly Occurring Periods of Eye Mobility and Concomitant Phenomena, During Sleep. *Science*, 118, 273–274.

Brand, A. and Eagleman, D. (2017). *The Runaway Species: How Human Creativity Remakes the World*. Edinburgh: Canongate Books.

Censor, N., Karni, A. and Sagi, D. (2006). A Link between Perceptual Learning, Adaptation and Sleep. *Vision Research*, 46(1), 4071–4074.

Charles, R. (2004). *Brother Ray: Ray Charles' Own Story*. Boston: Da Capo Press.

Cotrina, M. L. and Nedergaard, M. (2012). Brain Connexins in Demyelinating Diseases: Therapeutic Potential of Glial Targets. *Brain Resolution*, 1487, 61–68.

Darwall-Smith, H. (2021). *The Science of Sleep: Stop Chasing a Good Night's Sleep and Let It Find You*. London: Dorling Kindersley Limited (DK). Penguin Random House.

Foster, R. (2022). *Life Time: The New Science of the Body Clock and How It Can Revolutionise Your Sleep and Health*. Oxon: Penguin Life.

Guadiana, N. and Okashima, T. L. (2021). The Effects of Sleep Deprivation on College Students. *Nursing*, Senior Thesis 30. DOI: https://doi.org/10.33015/dominican.edu/2021.NURS.ST.09

Hassabis, D. (2016). Cited in Huffington, A. (2016). *The Sleep Revolution: Transforming Your Life, One Night at a Time*. London: Ebury Publishing.

Hauri, P. and Linde, S. (1996). *No More Sleepless Nights*. New York: John Wiley and Sons.

Hobson, A. (2005). Sleep Is of the Brain, by the Brain and for the Brain. *Nature*, 437, 1254–1256.

Huffington, A. (2016). *The Sleep Revolution: Transforming Your Life, One Night at a Time*. London: Ebury Publishing.

Kuriyama, K., Stickgold, R. and Walker, M. (2004). Sleep-Dependent Learning and Motor-Skill Complexity. *Learning and Memory*, 11(6), 705–713.

Leader, D. (2019). *Why Can't We Sleep?* London: Penguin Random House.
Littlehales, N. (2016), *Sleep: The Myth of 8 Hours, the Power of Naps ... and the New Plan to Recharge Your Body and Min*d. London: Penguin Random House.
Medina, J. (2008). *Brain Rules: 12 Principles for Surviving and Thriving at Work, Home and School.* Seattle, Washington: Pear Press.
Mignot, E. (2008). Why We Sleep: The Temporal Organization of Recovery. *PLoS Biology*, 29, 6(4) 0661-0669.
Nedergaard, M. (2013). Sleep Drives: Metabolite Clearance from the Adult Brain. *Science*, 342, 373–377.
Rechtschaffen, A. (1971). The Control of Sleep. In W. A. Hunt (Ed.), *Human Behavior and Its Control.* Cambridge, MA: Shenkman Press.
Smolensky, M. and Lamberg, L. (2001). *The Body Clock Guide to Better Health: How to Use Your Body's Natural Clock to Fight Illness and Achieve Maximum Health.* New York: Henry Holt Paperbacks..
Willett, A. and Barnett, J. (2017). *How Much Brain Do We Really Need?* London: Robinson.

CHAPTER 13

Why Targets Focus the Brain

'Set goals that stretch you but do not break you'. (Kelly 2007: 62)

There are many fallacies about the human brain. One is that it is like a sponge, absorbing everything and storing it somewhere. It is true that our brain probably looks and feels like a sponge, but it never acts like one. As we grow we re-organise our brain to cope with the demands we place on it. Millions of neural connections that are not used get pruned away. If we were to attend to every bit of data that hits our senses, we would not be able to cope with life. Selection is part of what the brain does to keep us healthy.

Defining our goals can mean organising our priorities, putting us on a narrow path to focus on what we want to achieve. The brain has a system that enables us to regulate different states of attention and alertness. Thus the process of selecting a goal or a target activates our brain's reticular activation system (RAS). In the first instance, the RAS acts as a filter for all the thoughts coming at us as we begin to explore options. It takes in all the thousands of sights and sounds that our brains perceive on a moment-to-moment basis. It then condenses them down into a smaller amount, so our conscious awareness can process them into something that is meaningful and useful.

At this point the amygdala and the frontal lobe work together to push us towards the completion of the goal. The amygdala, which is the brain's emotional centre, evaluates how important the information is to us. The frontal lobe, which is the brain's logical problem-solving part, defines the goal and digests it. Within the context of educational goals, a classic example of this is completing an essay by breaking it down into manageable chunks by using the concept of mind-mapping developed by Buzan (2005) and constructing each section within a time frame. When the goal is set and accomplished, the brain experiences a surge of dopamine, creating a sense of pleasure and satisfaction. This positive reinforcement encourages us to repeat the behaviour and take consistent steps towards objectives.

The distinction between a goal and a target can sometimes be hard to distinguish, because the two terms are often used interchangeably but have distinct meanings. A goal is a broad outcome that an individual aims to achieve. It is a desired result that guides decision-making and actions. Goals are usually long-term and can encompass multiple targets and usually require a plan. On the other hand, a target is a specific aim or objective that an individual wants to achieve. It is a measurable result that can be worked towards. Targets are usually short-term and they can serve as a benchmark for evaluating performance. To reach a goal means overcoming obstacles through persistence and effort in reaching a set number of targets. A practical example of the difference between a target and a goal is as follows; acquiring a recognised university qualification is a goal while successfully completing an assignment is a target.

Before we specifically look at the brain health benefits of setting targets, we will first outline the link between targets, the brain and dopamine. We will then examine the idea of mindsets and the Kaizen philosophy as a method towards achieving small victories, one at a time. Secondly knowing what to do to improve our brain health is different to knowing how to sustain motivation. For this we will examine the important role of self-efficacy and confidence, two strategies we can use based on neuroscience research.

The Brain, Targets and Dopamine

Setting targets literally alters the structure of the brain so that we perceive and behave in ways that will cause us to achieve those targets. Research on neuroplasticity suggest that our brain's structure adapts to support our targets (Berkman 2018). Achieving our targets is not just about hard work and discipline. It is about understanding how the brain processes success and failure. The more time we succeed with a target, the longer our brain stores the information that allowed us to do so well in the first place. That is because with each success our brain releases a chemical called dopamine.

Dopamine is a neurotransmitter that deals with how we feel pleasure. It is a motivation factor in our brain's reward system. It does not make us feel good in and of itself, but it does make us more motivated to chase down the things that do make us feel good (Medina 2008). It plays a key role in keeping us focused on our goals and motivating us to attain them, rewarding our attention and achievement by elevating our mood. We feel a sense of purpose, inner harmony and tend to pay more attention to our surroundings when we work towards a goal with measurable targets: for example, registering on a twelve-week module, attending classes and successfully completing assignments.

When dopamine flows in the brain's reward pathway (the part of the brain responsible for pleasure, learning and motivation), we not only feel greater concentration, but are inspired to re-experience the activity that caused the chemical release in the first place. Dopamine secretion gives a person a sense of rewarding pleasure and an energy boost. Its release also reinforces the connections among the neurons in the very network that helped us to perform the rewarding activity.

When it comes to setting personal targets for enhancing learning, the brain loves a challenge, generated through rewards such as a sense of achievement and subsequent happiness. According to Professor Ian Robertson (2022: 6), the more we do things that push us slightly outside our comfort zone, the more our confidence grows and the less anxiety we will feel the next time a similar challenge arrives. Whenever we achieve a goal that we have set for ourselves, it feels like a sense of accomplishment, that sense of satisfaction we get the activation of the reward network.

> **YouTube Video**
>
> In this video on situational vs well-developed interest, Dr Althea Kaminske describes the difference between situational and well-developed interest and the implications for learning. We don't have to be interested to be able to learn something, but it helps if the teacher can facilitate a move from situational interest (fleeting), to sustained interest, moderate personal interest, to well-developed interest. Project based learning, multi-modal learning strategies, spaced learning, concrete examples, elaboration can help students to develop their interests.
>
> Available from: <https://www.youtube.com/watch?v=FTe_NPq93FY> (10 minutes)

Deciding on a Target

Making decisions is a crucial part of the target-setting process. For many years, psychologists assumed that decision-making was a largely logical process. However we now know that very little of our decision-making is purely logical because emotions, prejudices, age profile and social interaction are very much influential in our decision making.

Within the brain when we make a decision to set a goal and plan, the part of the frontal lobe known as the orbitofrontal cortex, becomes active. Another area, the medial prefrontal cortex, becomes active when there is ambiguity or uncertainty involved in the choices we are trying to make. These lobes have been described as the 'control panel' of our personality, giving us the ability to make choices, to control our reactions and emotions, to project possible scenarios and action sequences and to use our imaginations. But the frontal lobes do not act on their own: their strong links with the amygdala, the parietal lobe and other brain areas make sure that our thinking is informed by our emotions, our past experiences and our stored knowledge. As we have pointed out in Chapters 7 and 8, this all combines to produce our experience of learning in general, and decision-making in particular.

In *Hare Brain, Tortoise Mind*, Claxton (1998) also describes the various ways of decision-making, highlighting the many ways that our thinking can be automatic or guided by our wits. 'Wits' for example are responsible for jamming on the breaks if someone pulls out in front of us in a car – there's no time to think as we react first and we think about what has happened afterwards. The second way we think is what Claxton termed the intelligent conscious. This is where we actively go about solving problems – how can I get to work in the least amount of time? The intelligent conscious is interested in solving the problem. Explaining and coming up with solutions to the situation are more important than simply observing it. Teaching and learning encourage us to adopt an intelligent conscious approach very well, while the quirkier, unconscious methods of problem-solving are often ignored.

The third form of thinking again according to Claxton, is our intelligent unconscious. The intelligent unconscious is more interested in the problem than necessarily finding a solution. It tends to be imaginative and playful, meandering around a particular topic or issue, 'messing about' to get to an answer. This mode of thinking is more useful when dealing with complicated or thorny issues. It is here that our most creative ideas occur. While the process of reaching into our subconscious may not be to everyone's taste, being able to use this very different way of thinking can be remarkably useful – especially when our normal, quick, solution-orientated approach doesn't seem to be generating any answers.

Staying Focused on Small Steps

A large part of success in anything is overcoming difficulties and experiencing small victories. Fundamentally succeeding in achieving any aspiration is mostly about overcoming our own self-doubt and fear of failure. Big aspirations or goals may seem a long way off with many months or even years of hard work between us and our objective. In some cases, this is not an encouraging feeling and can leave us wondering if it is really all worthwhile. Brain researchers tell us that the mental 'trick' is to learn how to enjoy the smaller triumphs – the stepping stones that take us towards our goal, in their own right.

Amabile, in her article *The Power of Small Wins*, presented a helpful concept which is known as the 'progress principle'. Over a decade of research, charting nearly 12,000 diary entries, she and her colleagues found that by far the biggest motivator for people was feeling that they were making meaningful progress in achieving their work goals. The most interesting thing was those moments of progress did not have to be spectacular- just feeling that the participants in the study were doing or providing something useful felt meaningful. A feeling of progress can create a positive feedback loop in the brain which leads to further progress being made.

Kaizen philosophy, pronounced Kaizen (Ky'zen), is used to mean 'continuous improvement'. It was originally designed as a business management technique to improve innovation. The Kaizen method asked workers to make one small incremental change to parts of their work routines (Harvey 2019). This had a profound impact on the work 'culture' in Japan, especially in the car industry (it is also known as the Toyota change management method). The basic principle of involving people in having some control and say in their work practice has been adopted (in many different guises) in nearly all modern industrial societies.

The Kaizen philosophy is simple to comprehend but rather more challenging to apply in practice. One of the best ways to achieve a target is for us to get our mind off the huge task in front of us and focus on a single action one step at a time. This calls for having a strategy to identify what is our target and then plan out the steps necessary to achieve our goal. Once a habit has been formed as a neural pathway in the brain, it becomes fixed as a routine, reinforced in our behaviour and part of our identity and who we are.

An applied example of the Kaizen philosophy might be to approach academic writing by using mind mapping. Mind mapping is a powerful way to visually organise our thoughts and ideas (Buzan 2005). It is a diagram that starts with a central idea or topic and branches out to include related concepts and details. Mind maps are useful for brainstorming, planning and clarifying our thinking in preparation for writing an academic essay. There are essentially four steps: (1) Construct a mind map by drawing an oval shape in the middle of the page. Write within it a topic, idea or concept in the form of a question. This acts as the nucleus from which other concepts will radiate (2) Branching out from the central idea, adding related thoughts, key words or sub-topics to form sub-branches, using key words (3) Expand sub-branches off the main branches into sub-topics and (4) Use lines to connect these elements, creating a web of associations (Buzan 2005).

By using mind maps learners can make associations between concepts, and by thinking more holistically, the learner can reinforce the links between concepts in the brain. This will result in the transfer of information from short-term to long term memory, both perspectives of thought needing to come together.

Confident Competent Mindsets

Everybody wants to be competent and confident and to do well at what they do. Self-efficacy is the belief that we can mobilise all our personal resources into effective courses of action to produce and regulate events in our lives. Our mindset is thus largely within our control and can manifest itself as a disposition for learning and progress, or an obstacle and mental block.

Stanford University psychologist Carol Dweck proposes that there are two mindsets we can bring to setting and achieving targets. According to Dweck, two mindsets fixed, or growth, mindset play an important role in all aspects of a person's life. She argues that the growth mindset allows a person to live a less stressful and more successful life through setting realistic targets that can be achieved through their own efforts.

From her research on how students experience learning, Dweck found that in students with a fixed mindset, they believed that their basic abilities, their intelligence and their talents are fixed traits. They have a certain amount but that's it. However in a growth mindset, students understand that their talents and abilities can be developed through effort, resilience and persistence. They do not think that everyone can be a genius, but they believe everyone can get smarter if they work at it. Fixed-mindset individuals dread failure because it is a negative statement on their basic abilities, while growth-mindset individuals do not mind or fear failure as much, as they realise that performance can be improved and that learning can indeed come from failure (Dweck 2008: 6).

Dweck argues that helping children understand how their brain works gives them a sense of control over their own learning but more importantly how to set learning targets for themselves. Furthermore, having learning targets are a powerful tool for students to understand what they are expected to learn and how they can achieve it. When students are clear on expectations of learning they tend to double their rate of learning. Moreover, when students are clear on expectations, they have a better chance of assessing their current performance and using feedback accurately. Learning targets help students grow by informing them exactly what they need to learn, which in turn enhances their achievement. Targets provide added motivation to students, as they feel highly motivated to reach the next level in their learning. Learning targets also help students take personal responsibility for their learning, which is an important lifelong learning skill to develop.

Attitudes, beliefs and mindsets matter for school and life success. Kirschner and Hendrick (2020) argue that generic claims like 'challenges should be embraced' can be harmful. It really depends on what kind of challenge and where we're at as a learner. Instead they suggest that self-efficacy may be a better lens through which to view learner mindset or disposition. They argue that self-efficacy is how well a person thinks they can deal with a specific challenge. Some learners with similar abilities can perform differently depending upon their self-efficacy. The more success a person experiences in a particular domain (e.g. Maths), the more self-efficacy they develop in that domain. The second most important factor in the development of self-efficacy amongst learners is whether they attribute success to internal factors (their own performance or ability) or external factors (luck or chance). When learners experience success or failure and attribute causes to those outcomes, those attributions in turn affect their expectation of future failure of success (Neelen and Kirschner, 2020).

> **Blog Posts**
>
> In this Learning Scientists Blog, Cindy Nebel asks *should we give up on growth mindset?* She provides a list of resources which question the results of growth mindset interventions, and create a conversation about alternative theories and concepts which deal with learner disposition and performance success. Available from: <https://www.learningscientists.org/blog/2022/11/17/digest-167>
>
> Similarly, Mirjam Neelen and Paul A. Kirschner (2020) wrote a blog entitled – *Goodbye Growth Mindset, Hello Efficacy and Attribution Theory*. That learners 'think differently about competence and intelligence is true, but the influence of these ways of thinking on learning performance has never been well demonstrated' (Kirschner and Hendrick 2020: 91). They argue that self-efficacy and attribution theory are alternative ways to view learner dispositions, how they impact performance. Available from: <https://3starlearningexperiences.wordpress.com/2020/06/02/goodbye-growth-mindset-hello-efficacy-and-attribution-theory/>

Strategies for Target Setting

The 80/20 rule is one of the most useful of all concepts of time and life management. It is also called the 'Pareto Principle' after its founder, the Italian economist Vilfredo Pareto, who first wrote about it in 1895. Pareto noticed that people seem to divide naturally between those who achieve their goals and targets, and those who want to but do not. Basically the rule says that 20 per cent of our activity will account for 80 per cent of our results as this percentage has been prioritised. This means that if we have a list of ten items to do, two of those items will turn out to be worth as much if not more than the others. The sad fact is that most people do not discriminate between the significant and the more trivial items resulting in a lack of efficiency in choosing priorities.

In this context keeping a goal in our head is not nearly as effective as writing it down. Researchers in the area of health psychology have demonstrated that writing things down serves as a 'focusing device'. It also increases our commitment to turn an abstract long-term goal into short-term reminders. It's like the phrase 'Don't count the days. Make every day count'. Writing things down helps create an inner dialogue which emphasises the importance of what we are doing. The target should be realistic and specific, and incorporate our strategy. Writing it down will help us stay focused, harnessing the power of the brain's reward centre. Here for some people keeping a diary lends an objective power to what they are trying to achieve.

Each brain health learning target we create should be SMART: specific, measurable, attainable, relevant and timed, with a deadline. You may have heard the expression 'the secret of getting ahead is getting started'. True but neuroscientists warn against setting unrealistic goals – the bigger the goal puts the brain under more stress. Setting the bar too high can actually be counterproductive. Every time we fail the brain is drained of dopamine, making it not only hard to concentrate but also difficult to learn from what went wrong. Limit the goal to a small number of reasonably attainable objectives.

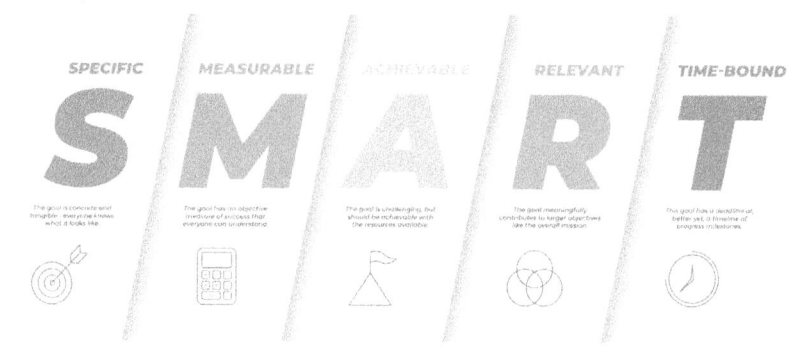

Figure 18: SMART Targets – Image by *Diki*/<stock.adobe.com>

Let's take a moment to explore the concept of a SMART target:

Specific: The target must be clearly defined with parameters that state what, when, who, where and how. Specific targets help us stay focused on one thing at a time. For example, stating that we want to become a member of a tennis club to get regular exercise is fine, but we can define our target more clearly by saying you want to become a member of a particular club and be involved in a regular doubles game every Tuesday evening to 7 p.m. or 8 p.m.

Measurable: The target must include a way of measuring the result. This typically means that there is a number associated with the target, such as how many or what percentage – some quantifiable way to measure progress toward the target. Stating that we want to write a book is neither specific nor measurable; it's more useful to say specifically that we will write 200 words a day for five days of the week – the target is to write 1,000 words a week and 4,000 words a month. Do not be too rigid, be flexible. Above all be kind to ourselves. If something unexpected turns up this week to prevent us achieving a mini-target, there is always next week.

Attainable: Each target that we create must be within our reach. It should not be so far-fetched that it is out of our vision range to determine whether a target is attainable consider some of the practical realities – what other pressures are coming up next year that may impede or improve our ability to reach the target? Also consider what needs to be done to accomplish the target, especially the time required and the motivation to keep going. Any change or any development of a new routine will take hard work, dedication and focused attention.

Relevant: The target must have relevance and meaning for us, otherwise, we will not be motivated to accomplish it. What will be the outcome if we meet a particular target? Will it add to the quality of our life, acquire a new skill, considerably enhance our health, make more money, improve our career prospects, save us time in the long run? Whatever our personal motivation, ensure that targets tap into it. Doing so will inspire us every day to keep striving toward the finishing line.

Timed: Speaking of a finishing line, the target must have a deadline or completion date assigned to it. Without one we will lose our desire to meet the target in a timely manner. Targets may be specific and measurable, but how long will it take and when will we measure them?

Whatever learning brain health target we aim to achieve, remember to develop the habit to always include it as an item in the top 20 per cent of targets. As we have seen, when we actually begin work on a valuable task, we release dopamine in our brain, which in turn increases motivation to continue. Our ability to discriminate between the important and unimportant is a key determinant of working towards our targets that can really make a difference in our life. As a result you accomplish vastly more than the average person and are much happier as a result.

Conclusion

In this chapter we have explored how setting targets has a powerful impact on the human brain as well as triggering the brain's own healing system through the release of the neurotransmitter dopamine as well as different hormones and endorphins. The brain's capacity to change through neuroplasticity creates the possibility of achieving the best possible brain health as we age. Learning to develop strategies for setting and reaching targets means that neurons can make the new connections needed to learn new things. This builds cognitive flexibility and cognitive reserves.

Building cognitive flexibility and cognitive reserve means that all the parts of the brain work together in harmony as the process never involves just one structure alone. As pointed out in earlier chapters the basal ganglia do this by ensuring learning gets built into the neural forest of the cerebral cortex. They also smooth out fine motor action together with the motor context of the brain which help us retain learning as habits in the cerebellum as well as clarifying our thoughts. We have used examples in this chapter of mind mapping, SMART planning and Kaizen philosophy to illustrate how learners can work with targets to consolidate learning.

References

Amabile, T. and Kramer, S. (2011). *The Power of Small Steps.* New York: May Issue Harvard Business Review.
Berkman, E. (2018). The Neuroscience of Goals and Behavior Change. *Consulting Psychology Journal*, Mar; 70(1), 28–44.
Buzan, T. (2005). *The Ultimate Book on Mind Maps.* London: Thorsons.
Claxton, G. (1998). *Hare Brain, Tortoise Mind: Why Intelligence Increases When You Think Less.* London: Fourth Estate.
Damasio, A. (2008). *'Descartes' Error: Emotion, Reason and the Human Brain.* New York: Random House.
Dweck, C. (2008). *Mindset: The New Psychology of Success.* New York: Ballantine Books.
Harvey, S. (2019). *Kaizen: The Japanese Method for Transforming Habits One Small Step at a Time.* London: Pan Macmillan.
Hayes, N. (2018). *Your Brain and You: A Simple Guide to Neuropsychology.* London: Teach Yourself. Hodder and Stoughton.
Kahneman, D. (2011). *Thinking, Fast and Slow.* London: Farrar, Straus and Giroux.
Kelly, M. (1999). *The Rhythm of Life: Living Everyday with Passion and Purpose.* Sydney: Barnes and Noble.
Kirschner, P. and Hendrick, C. (2020). *How Learning Happens: Seminal Works in Educational Psychology and What They Mean in Practice.* London: Routledge.
Medina, J. (2008). *Brain Rules: 12 Principles for Surviving and Thriving at Work, Home and School.* Seattle: Pear Press.
Neelan, M. and Kirschner, P. A. (2020). Goodbye Growth Mindset, Hello Efficacy and Attribution Theory. *Three Star Learning Experiences Blog.* Available from: <https://3starlearningexperiences.wordpress.com/2020/06/02/goodbye-growth-mindset-hello-efficacy-and-attribution-theory/>
Pareto, V. (1963). *The Mind and Society: A Treatise on General Sociology.* New York: Dover.
Robertson, I. (2022). Essential Guide to Ageing Well. *Age Action Ireland Magazine*, May.

CHAPTER 14

Yearning for Meaning

> 'Excellence is a continuous process and not an accident'.
>
> (Kalam, A. P. J. Abdul 2003)

Yearning is a strong desire or longing for something and it at the core of being human. It can be a powerful motivator for learning as it drives individuals to seek out knowledge and skills that will help them achieve clear goals. When someone yearns to learn they are likely to be engaged and interested in the learning process, which can lead to better retention and more effective learning outcomes.

Everyone experiences yearning; we ruminate, we wish, we dream. We may not talk about our yearnings and we may not allow ourselves to think about them often, especially if the things for which we yearn are far-away or unattainable for now. But our yearnings are integral to our brain health and who we are. Godwin (n.d.) states 'the best antidote I have found is to yearn for something. If you yearn, you cannot congeal, there is a forward motion to yearning'. Having a sense of purpose is the capstone to a life of health and healthy relationships. As Lillian Gish (n.d.) once quipped 'a happy life is one spent in learning, earning and yearning'.

Yearning differs from other ways of learning and knowing. Strictly speaking it is an act of the intellect commanded by the will. As well as intellect and will, human motivation is central to understanding the experience of yearning. For this reason our yearnings are intimately connected to our mind set, resilience, motivation, sense of self and belonging, our hopes for ourselves and our future goals. The most important questions we face are related to our yearnings on what matters in our lives? How can we live well in the world? How can we grow and achieve our goals?

Yearning for learning is a natural human instinct that drives us to seek knowledge, understanding and betterment. The joy in learning comes just as much from what is learned as from the learning process itself. In this chapter we will place yearning as a legitimate and crucial aspiration for both brain health and education. Specifically we will describe and explore the human urge for yearning as a foundation for building resilience within the framework of lifelong learning. We will discuss how pursuing one's goals, embracing an open mindset and maintaining motivation for growth, and learning to adapt to challenges as they arise is a fundamental part of success. We advocate for educational experiences that gift us opportunities to embrace a sense of identity, belonging, autonomy, aspiration, culture and allow us to fulfil our yearning for meaning.

Yearning as the Foundation for Resilience

Humans possess an innate capacity to strive for health and well-being, to bounce back from setbacks and to keep going in the face of tough demands and difficult circumstances (Seligman 2009). Resilience is this capacity 'to navigate their physical and social ecologies to provide resources, as well as their access to families and communities who can culturally navigate for them' (Ungar, Brown, Liebenberg, Cheung and Levine 2008: 168). Resilience isn't about being 'tough', it is about having emotional strength, awareness and a positive outlook on life. The more proactive and responsive we are in drawing on supports (both internal and external) when navigating stress, challenges and adversity, the more resilient we become.

We live in a society where stress is ubiquitous but not all individuals have negative health outcomes in response to stress. Instead some people achieve optimum health despite their exposure to potentially disabling stress factors. This observation was examined by Aaron Antonovsky, a professor of Medical Sociology in the 1980s. As part of a research project in Israel, he interviewed women about how they perceived their health, and about various life events affecting them such as the loss of a husband or partner, losing their eyesight or having suffered a serious illness. After analysing the interview results he found that 29 per cent of the women reported good health, even though this particular group of women survived the Holocaust. This finding intrigued him because how come some people achieve positive health despite their exposure to extreme hardship and stress. This led him to focus on this small number of respondents and a search for their health resourcefulness despite the psychological impact of surviving concentration camps.

In *Unravelling the Mystery of Health* (1987) Antonovsky asks why do some people, regardless of major stressful situations and severe hardship, stay healthy and others do not. To answer this question, he coined the term 'Salutogenesis' – a positive health model as he wanted to move away from the traditional pathogenesis model which focused on factors that cause disease. In contrast he was concerned with the relationship between health, stress and coping and with wider life situations or life orientation factors that have an impact on health. He wanted to focus on strength and motivation as a way to maintain and improve health. To better understand what it is that keeps people healthy he uses the sense of coherence concept. A sense of coherence refers to an enduring attitude about how people view life and when faced with stressful situations how they use their general resistance resources to maintain and develop their health.

What is interesting about this approach is that it does not postulate a sense of control as a primary characteristic of a healthy personality. Rather Antonovsky (1996) refers to a 'sense of coherence' (SOC) which is central to successful coping with challenge. He described it as a personality trait that expresses an individual's complete outlook on life. The sense of coherence consists of three components that are dynamically interconnected:

1. Comprehensibility (understanding the challenge, in other words the extent to which a person perceives that the world makes sense and is structured, consistent and predictable).
2. Manageability (believing in the availability of coping resources – the degree to which a person has sufficient resources at their disposal to meet both external demands and internal needs).

3. Meaningfulness (in terms of wanting to cope or the degree to which a person feels that life is a challenging but worthy emotional engagement).

The above model is consistent with the World Health Organisation's (WHO) approach to health. The WHO defines health as 'a state of complete physical, mental and social well-being and not merely the absence of disease or infirmity' (WHO 2021: 2). An important implication of this definition, according to the WHO is that mental health is more than just the absence of mental disorders or disabilities.

It turns out that optimistic dispositions are critical in developing a positive lifestyle. Winston Churchill is credited with the saying 'A pessimist sees the difficulty in every opportunity. An optimist sees the opportunity in every difficulty'. In *Rainy Brain, Sunny Brain,* neuroscientist Elaine Fox explores the little-understood connection between optimism and happiness, showing how we can brighten our lives by retraining our brain (Fox 2012). For instance, the Italian author Primo Levi (1986) recounts his experiences as a young chemist from Turin in a German concentration camp during the Second World War. In unsentimental language he chronicles the horrific story of a year at Auschwitz. However he never seemed to lose sight of the resilience of the human spirit. He attributed his survival largely to his capacity to perceive his fellow inmates, as well as himself, as people and not as objects. Holding on to this perspective allowed him to avoid the demoralisation or what he called the 'spiritual shipwreck' that engulfed so many others. In a later book, he describes his long trek to freedom marching across Eastern Europe and Russia, where 'vigorous people full of the love of life' rekindled in him the joy of living that the camps had almost extinguished. His account gradually unfolds as a story of hope, echoing the experiences of many who have come through great adversity.

While personal strength and character are central to resilience, there is a growing body of research that looks at the external or protective factors around individuals who appear resilient (Luthar at al. 2006; McGraw et al. 2008 cited in Worsley 2010). Having conducted a literature review into various studies on young people facing adversity,

Bernard (2004) found that resilience is a capacity that all young people have for health development and successful learning. Teachers can create the nurturing and empowering environments that in turn engage learners' innate resilience by developing their capacities for positive development, belonging and connectedness (Benard and Slade 2009).

Seligman suggests that resilience is 'the ability for an individual to bounce back from setbacks, and to keep going in the face of tough demands and difficult circumstances' (2009: 19). It is emotional strength and awareness and a positive outlook on life. We can promote human strength in the face of adversity through simple and practical actions – by recognising our emotions, findings social support, setting realistic goals, monitoring our experience, self-regulating and exercising positive self-talk (Seligman 2002). Lyn Worsley in her book *The Resilience Doughnut* says that adults tend to focus too much on young people's problems. She argues that we can learn much about young people by asking what makes them resilient despite the adversity they are facing. She proposes a model of a doughnut divided into 7 factors which contribute to young people's resilience. They are (1) Family factor (2) Parent factor (3) Education factor (4) Skill factor (5) Peer factor (6) Community factor (7) Employment factor. Most young people have some but not all factors working in their lives. Worsley's research found that resilient young people need a minimum of 3 of these factors. Furthermore. if young people focus on such factors their mindset has the potential to develop further in this context.

Building on Seligman's work (2009), the diagram overleaf highlights the many factors contributing to resilience. These factors could be integrated into a resilience building curriculum. We believe that such a programme could be developed and implemented during the transition year in all secondary schools as well as building on the concept of neuroplasticity as proposed by educational neuroscientists.

Factors contributing to Resilience:
- Optimism
- Self-belief
- Control of Self
- Willingness to Adapt
- Willingness to Be Flexible
- Ability to Solve Problems
- Emotional Awareness
- Social Support
- Sense of Humor

Figure 19: Factors Contributing to Resilience – Image by Dmitry/<stock.adobe.com>

Yearning for Motivation

Psychologist Abraham Maslow (1962) believed that human motivation is based on people seeking fulfilment and change through personal growth. His hierarchy of needs is a motivational theory in psychology comprising a five-tier model of human needs, often depicted as hierarchical levels within a pyramid. Needs lower down in the hierarchy must be satisfied before individuals can attend to needs higher up. From the bottom of the hierarchy upwards the needs are physiological, safety, love and belonging, esteem and self-actualisation. Maslow would argue that as people have their lower needs satisfied, they tend to move on to satisfy the higher ones.

Instead of studying neurotics, as Freud and Jung had done, Maslow decided to study people who were successful in life, to see what made them happy and content. He called them 'self-actualisers'. Self-actualisation he wrote, may be loosely described as the full use and exploitation of talents, capabilities and opportunities. Such people seem to be fulfilling themselves and to be doing the best that they can do. For Maslow self-actualisation refers to a person's yearning for self-fulfilment. Self-actualised people are those who are fulfilled.

Maslow argued that a person never remains static and is always searching for meaning. By self-actualisation a person comes to find a meaning in life that is important to them. They may seem to throw themselves into a cause and strive to obtain a meaningful goal. He found that self-actualisers were different from other people in a few ways. Whereas most people tended to be self-centred, and mistrustful of threatening reality, self-actualisers were self-reflective and open to the mysteries of the world. Whether they were believers or non-believers in a religious organisational sense, self-actualised people were usually spiritual in the sense that they often reported having peek-experiences of a mystical kind. Maslow believed self-actualisation could be measured through this rather ambiguous concept of peak experiences.

Another psychologist Mihaly Csikszentmihalyi (1997), calls this type of engagement 'optimal experience' or 'flow'. This describes a state of complete engagement with a task to the point where we lose track of time and want to continue working on the task as long as possible. In this moment there is no sense of past, no sense of future, just an intense present, an overwhelming feeling of being 'in the moment', or what sports people call 'in the zone'. This is the magical moment when the mental and physical merge in an effortless way – that is, when the basal ganglia works to ensure efficient automatic movement that is driven by practice. For example, if you have ever forgotten to eat lunch because you were so absorbed in finishing a task, you have experienced a 'flow' state.

According to Csikszentmihalyi's research, most of us experience this type of flow about once every couple of months. Almost 12 per cent of people say they never experience this sensation, while 10 per cent say they experience it daily. It appears this state happens when we have clear goals, forget our personal problems and merge with the actions we are performing. The trick is to find the right balance between our level of skill and the degree of challenge. If the task is too easy, boredom is the most likely outcome, if too hard it becomes stressful. But when the difficult level presents us with a genuine but realistic challenge, then a trancelike state where nothing else matters can be entered.

We know that learners perform better during higher order cognitive tasks such as problem solving, critical thinking and creativity; these tasks tend to be driven by intrinsic motivation that is, the internal desire for more information or 'the answer' to a solution. However motivation for these tasks can quickly become extrinsic by providing grades as the reward or outcome for the task (Pink 2011). External motivation will ensure short-term accomplishments but it can only take us so far. In the child's world we regularly provide extrinsic motivation through rewards, grades, competitions, prizes, gold stars and praise at school level but in the adult world most of these rewards are frequently lacking. We need to prepare learners with a strong sense of intrinsic motivation who are more likely to adopt a lifelong learning mindset.

Herman argues that 'in order to maximize the effectiveness of our pedagogies, we must understand how our pedagogies align with prevailing theories of cognition and motivation and design our pedagogies according to this understanding' (2012: 369). Connecting lessons or tasks to prior learning or tailoring it to the interests of the learner often leads to a deeper understanding of the practical applications of knowledge. Finding a learner's passion is the key to identifying that which 'they find important, and in which they invest time and energy' (Vallerand et al. 2003: 756). Learners are more likely to see the relevance of what they are learning to real-world situations if they make their own connections which in turn can enhance their motivation and engagement. They view learning as a continuous and enjoyable process rather than a task to be completed.

Neuroscience research has shown us that stimulating curiosity in learners, and the added experience of novelty and surprise, can enhance learning and memory (Oudeyer 2016). In a study mixing behavioural analysis and brain imaging, Kang et al. (2009) showed that adults show greater long-term memory retention for verbal material for which they had expressed high curiosity. Within the brain the motivational systems that signal the value of primary rewards (dopamine) are activated by the desire to obtain information. Therefore curiosity-driven learning and intrinsic motivation are fundamental ingredients for deep learning (Freeman et al. 2014). Learners who are intrinsically motivated are more likely to invest time and effort into mastering a skill or understanding a concept, even when faced with obstacles. In a school context, learners who have a sense of self-efficacy enjoy autonomy when they make decisions about their learning, and gain a sense of purpose when the task enables them to pursue their personal goals (Pintrich 2003).

Our intrinsic motivation is influenced by our social and emotional lives. Relationships with parents, teachers and peers all play a role in young students' academic motivation (Ryan et al. 1994). Furrer and Skinner (2003) found that the relationship with a teacher (as discussed in Chapter 9) provided the most influence on learners' emotional engagement in the classroom, and was a good indicator of learners' motivation. There seems to be a tension between the need for autonomy and for social interaction but at the same time learners' relationships do provide a solid base for their autonomous action which can spark intrinsic motivation.

When learners feel a sense of control and choice in their actions, they can pursue activities aligned with their personal preferences and values. They are more likely to find pleasure and satisfaction in the process of the activity itself, rather than focusing solely on external outcomes or rewards. Intrinsic motivation plays a crucial role in the learning process as it fosters a willingness to persist in the face of challenges. Furthermore, the design of lessons and tasks that support intrinsic motivation depends on teachers' attitudes and class structures (Ryan and Deci, 2002). For instance, teachers who focus on the learning process over the learning outcome tend to ask reflective questions, provide less direction, design problem-based learning strategies and are more likely to support student's intrinsic motivation (Hernman et al. 2012; Herman 2012; Reeve and Halusic 2009).

Yearning for Happiness

Happiness and yearning are often linked, but it is important to remember that they are not the same thing. Happiness is largely how we feel here and now: the contentment we feel about how our life is going, or the joy we experience when we celebrate a happy family event. Yearning is how we think and feel about the future. If we really do believe that things will work out for the best, all setbacks become easier to deal with.

Brendan Kelly in his 2021 book *The Science of Happiness* shows that our lifetime's happiness follows a U-shaped curve, with the research consistently showing across cultures worldwide that forty-seven years is the bottom of the U, the age when most people are unsettled. But regardless of our age, our happiness is influenced by multiple factors including who we are, where we live, what we do and how we choose to live our lives. While we each have a genetic predisposition to a certain level of happiness, through a gentle but considered approach of healthy choices, we can all improve our day-to-day contentment.

In the modern world, our basic needs – food, shelter, warmth – are usually met, but it is the connectedness with others and a sense of meaning that are often missing. This is the root of what George Easterbrook (2003) calls the 'Paradox of Progress'. He found that in the United States and Europe, while the level of wealth grew substantially over a fifty-year period from the 1950s onward, happiness levels did not budge, and rates of anxiety, addiction and depression soared. Often there is a disconnection between the level of material wealth in a society and the subjective feelings of happiness and well-being. It also appears that a more egalitarian society, with higher welfare provision and the prioritisation of community, reports higher levels of happiness.

However survey after survey has found that genuine changes in perceived happiness only comes about when three things come together: lots of positive emotions and laughter; being fully engaged with our lives and finding a sense of meaning that is broader than our day-to-day life. Of these three components, engagement in what we are doing, whether it is work or leisure, seems to be especially important. What does make us happier is getting involved in something that is meaningful for us. Additionally, neuroscientists tell us that yearning for a better life is central to switching on brain circuits that home in on the motivational landscape in our brain. The activity in certain brain areas determines what we tune into and what we respond to. When our positive motivational brain becomes active it can lead to conditions conducive to happiness.

Psychologist Barbara Fredrickson (2009) is an expert on the concept of flourishing, which she broadly defines as living at the top of our range, enjoying our life filled with goodness, growth, and creativity, and, when things go wrong, a strong resilience to get over the hump. For many years she has been an advocate of finding ways to bring more positive emotions into our lives. In her research she discovered a critical 3:1 ratio, indicating that we need to have three positive emotions for every negative one to thrive. Positive personal emotions include things like a sense of wonder, compassion, contentment, gratitude, hope, joy, and sexual desire, while negative feelings are those like anger, contempt, disgust, embarrassment, fear, sadness, and shame, to name just a few.

Fredrickson has found that if we really want to prosper, we should not try to eliminate negative emotions; rather, we should work on keeping the ratio at three positives for every one negative. Most of us, she has found, have two positive experiences for every negative one. This gets us by, but it is only we can get this ratio up to three positives for every bad experience, we can begin to truly flourish. The 3:1 ratio is the critical dividing line between those who were getting the most out of life and those who were not. This research finding from psychology is also supported by studies investigating the link positive relationship between yearning and spirituality.

Conclusion

Yearning for learning is a natural human instinct that drives us to seek knowledge, understanding and betterment. The joy in learning comes just as much from what is learned as from the learning process itself. In this chapter, we have placed yearning as a legitimate and crucial aspiration for both life, brain health and education. We examined the notion that yearning is deeply human and encompasses a wide spectrum of meanings. Pursuing one's goals, embracing an open mindset and maintaining motivation for growth, and learning to adapt to challenges as they arise are fundamental parts of success. We described and explored the human urge for yearning as a foundation for building resilience within the framework of lifelong learning. We also advocate for educational experiences that gift us opportunities to embrace a sense of identity, belonging, autonomy, aspiration, culture and allow us to fulfil our yearning for meaning.

References

Antonovsky, A. (1987). *Unravelling the Mystery of Health*. San Francisco: Jossey-Bass.
Antonovsky, A. (1996). The Sense of Coherence: An Historical and Future Perspective. *Israel Journal of Medical Studies*, 32, 170–178.
Benard, B. (2004). *Resiliency: What We Have Learned*. San Francisco: Wested.
Csikszentmihalyi, M. (1997). *Finding Flow: The Psychology of Engagement with Everyday life*. California: University of California.
Easterbrook, G. (2003). *The Paradox of Progress: How Life Gets Better While People Feel Worse*. New York: Random House.
Fredrickson, B. (2009). *Positivity: Groundbreaking Research Reveals How to Embrace the Hidden Strength of Positive Emotions, Overcoming Negativity, and Thrive*. New York: Crown.
Freeman, S., Eddy, S. L., McDonough, M., Smith, M. K., Okoroafor, N., Jordt, H. and Wenderoth, M. P. (2014). Active Learning Increases Student Performance in Science, Engineering, and Mathematics. *PNAS*, 111(23), 8410–8415..
Furrer, C. and Skinner, E. (2003). Sense of Relatedness as a Factor in Children's Academic Engagement and Performance. *Journal of Educational Psychology*, 95(1), 148–162.
Galahena, T. (n.d.). A Sri Langan Poet, Yearning. Available from Poet Hunter.com.
Geoffrey, L. Herman. (2012). Designing Contributing Student Pedagogies to Promote Students' Intrinsic Motivation to Learn. *Computer Science Education*, 22(4), 369–388.
Herman, G. L., Trenshaw, K. and Rosu, L. (2012). Work-in-Progress: Empowering Teaching Assistants to Become Agents of Education Reform. In *Proceedings of the Forty-Second ASEE/IEEE Frontiers in Education Conference*. 3–6 October, Seattle, WA (pp. T2C-1 to T2C-2). Washington, DC: ASEE.
Kang, M. J., Hsu, M., Krajbich, I. M., Loewenstein, G., McClure, S. M., Wang, J. T. and Camerer, C. F. (2009). The Wick in the Candle of Learning: Epistemic Curiosity Activates Reward Circuitry and Enhances Memory. *Psychological Science*, 20(8), 963–973: 8410–8415.
Kelly, B. (2021). *The Science of Happiness*. Dublin: Gill Books.
Levi, P. (1986). *Surviving in Auschwitz and the Reawakening*. New York: Summit.
Luthar, S. S., Shoum, K. A. and Brown, P. J. (2006). Extracurricular Involvement among Affluent Youth: A Scapegoat for 'ubiquitous achievement pressures'? *Developmental Psychology*, 42(3), 583–597.
Maslow, A. (1962). *Motivation and Personality*. London: Penguin.

McGraw, K., Moore, S., Fuller, A. and Bates, G. (2008). Family, Peer and School Connectedness in Final Year Secondary School Students. *Australian Psychologist*, 43(March 2008), 27–37.

Oudeyer, P., Gottlieb, J. and Lopes, M. (2016). Intrinsic Motivation, Curiosity and Learning: Theory and Applications in Educational Technologies. *Progress in Brain Research*, 229, 257–284.

Pink, D. H. (2011). *Drive: The Surprising Truth about What Motivates Us*. New York: Riverhead Trade.

Pintrich, P. R. (2003). A Motivational Science Perspective on the Role of Student Motivation in Learning and Teaching Contexts. *Journal of Educational Psychology*, 95, 667–686.

Ryan, R. M. and Deci, E. L. (2000). Self-determination Theory and the Facilitation of Intrinsic Motivation, Social Development, and Well-being. *American Psychologist*, 55, 68–78.

Ryan, R. M., Stiller, J. D. and Lynch, J. H. (1994). Representations of Relationships to Teachers, Parents, and Friends as Predictors of Academic Motivation and Self-esteem. *Journal of Early Adolescence*, 14(2), 226–249.

Seligman, M. (2002). *Authentic Happiness: Using the New Positive Psychology to Realize Your Potential for Lasting Fulfilment*. New York: Free Press.

Ungar, M., Brown, M., Liebenberg, L., Cheung, M. and Levine, K. (2008). Distinguishing Differences in Pathways to Resilience among Canadian Youth. *Canadian Journal of Community Mental Health*, 27(1), 1–13.

Vallerand, R. J., Blanchard, C., Mageau, G. A., Koestner, R., Leonard, M., Gagne, M., et al. (2003). Les Passions de l'Ame: On Obsessive and Harmonious Passion. *Journal of Personality and Social Psychology*, 85(4), 756–767.

World Health Organisation. (2021). *Comprehensive Mental Health Action Plan 2013–2030*. Geneva, Switzerland: World Health Organisation (WHO).

Worsley, L. (2006). *The Resilience Doughnut: The Secret of Strong Kids*. Sydney: Wild and Woolley Publications.

CHAPTER 15

Liquids: Elixirs of Life

'The mind is like water, when it's turbulent, it's difficult to see. When it's calm, everything becomes clear'.

(Prasad Mahes cited in Anderson 2020)

Liquids are indispensable for life. We can survive for possibly weeks without food, but only a few days without water. Water is the most abundant liquid in the human body, making up about 70 per cent of our body weight. It performs a host of important internal functions, from maintaining body temperature, to transporting vitamins, hormones, and to lubricating joints, eyes, skin and intestines. It improves mental alertness, is critical for our ability to concentrate and learn and it needs to be replenished on a continuous basis.

Although it is well known that water is essential for human survival, only recently have we begun to understand its role in maintaining brain function. It is often overlooked as a significant nutrient that can affect, not only physical performance but also mental performance. When the body is low in water, it is difficult for the brain to focus on a task, pay attention and comprehend information. It is not really surprising then that regular supplies of water and other liquids are so essential to cognitive functioning – after all, while the brain weighs only 3 pounds, it uses 20 per cent of the body's energy and, in an adult of average weight, the body contains roughly 40 litres or seventy pints of water. Furthermore, the brain is 73 per cent water so even experiencing moderate dehydration by just 2 per cent can lead to headache, dizziness, fatigue and lethargy.

All the cells in our bodies, especially brain cells, depend heavily on water to carry out essential functions. Therefore, if our water levels are low, our brain cannot function properly which can lead to cognitive problems.

The source of food to the brain is through the blood. Blood brings nutrients to the cells and takes away toxins. A brain cannot survive without oxygen even for short periods of time, even when we sleep, without suffering damage and ultimately death. Our brain works around the clock and, to avoid thirst, requires more water than any other part of the body.

Thirst is one of the strongest drives the body possesses. If we are denied both food and water, we feel thirsty long before we feel hungry. Thirst is the body's way of telling us it needs to be hydrated. The amount of water our body needs depends on many factors, and our natural thirst mechanism kicks in long before we become dehydrated. Quenching our thirst is important for physical and mental health and has a key role in reducing fatigue.

In this chapter we will outline and discuss why our brain needs to stay hydrated because water: (1) Keeps the blood flowing properly and delivers oxygen, vitamins and minerals to our brain (2) Produces hormones and neurotransmitters and help reduce cortisol levels (3) Maintains the fluid levels that shield brain cells from shrinking and mild cognitive impairment (4) Maintains the process of neurogenesis (the growth of new neurons) and produce (increases function and longevity of all brain cells (5) Flushes the metabolic waste and toxins that build up in the brain, As well as the above there is the therapeutic use of water and the rejuvenation effects of increased perspiration.

A Healthy Brain Needs a Healthy Body

The first principle to maintaining a well-functioning brain is to maintain a healthy body. The primary purpose of every organ system of the body is to serve the brain. It is the nutrients in our food that provide the vital mental building blocks to enable learning. The key nutrients to sustain and promote mental and physical activity are water, proteins, carbohydrates, fats, vitamins and minerals. Whatever we eat and drink is absorbed into the blood which in turn dissolves and transports nutrients to the brain. New research suggests that brain cells do not age as fast as we thought. Rather it is the blood vessels that feed them that age. According to neuroscientist

Daniel G. Amen, 'whatever is good for the heart is good for the brain' and if we want to keep our brain healthy, it is critical to protect our blood flow (Amen 2016). Furthermore, he suggests that we 'focus on strengthening our blood vessels, by avoiding stress, as well as excessive caffeine and nicotine as these constrict blood flood to the brain and other organs). Poor blood flow is devastating to brain function, so drinking plenty of water is recommended' (2011: 359).

Water is a basic need for cellular health. The cells throughout our body and the neurons in our brain need water to function efficiently. Cells contain water and are surrounded by water. After all, approximately two-thirds of the body's water content is contained within each of the 100 billion or so neurons, and the remaining third is the fluids filling the spaces between the cells and the water in the bloodstream. Therefore more than any other cell in the human body, neurons require more water because thirst hampers the communication process between them, neurons as well as hampering the flow of neurotransmitters.

The Mechanism of Hydration and Thirst

Thirst occurs when water levels fall and the body is motivated to take in more water. This produces the urge for us to drink fluids, and thus thirst is an essential mechanism in fluid balance and ultimately is essential for survival. The craving to drink when we are feeling parched might feel intuitive, but the body and brain rely on an intricate set of biological processes to make sure we stay properly hydrated, as consuming both too little or too much water can lead to problems Hydration is therefore essential to protect and nurture cells throughout the body and brain.

There are different types and kinds of thirst and the body and brain respond to these kinds of thirst differently. Inside the forebrain, below the thalamus, is a particular organ called the hypothalamus which is responsible for regulating body temperature, thirst, hunger, sleep and other homeostatic systems among other functions? With special sensors on it, the hypothalamus continually monitors fluid concentration of sodium and other useful substances in our blood. This is how the brain interprets two different kinds of thirst: extracellular thirst and intracellular thirst.

Extracellular thirst is triggered by a decrease in the volume of essential fluids – which is the fluid outside the cells. This can occur due to loss of fluids from the body, such as when sweating, urinating or when feeling unwell through bleeding, vomiting or diarrhoea. In contrast, intracellular thirst is caused by a depletion of fluids within the cells, when concentration of salt is too high from consuming salty food. Another name for this is osmotic pressure, which is when fluids are forced out of the cell. However in both cases the hypothalamus sends us a strong message to drink water or eat fruits high in water content.

On average an adult handles roughly 3 litres of water per day between intake and output. Water enters and leaves the body by various routes controlled by the mechanism of thirst and regulatory hormones in the brain. Liquids enter the body in 3 forms: (1) water taken in as water or other beverages, (2) water in food and (3) as metabolic water produced by cell oxidation. In general roughly 70 per cent of fluid intake comes from water and other beverages, 20 per cent comes from food and 10 per cent through cell oxidation. Many common foods contain large amounts of water, such as bread, cereals and biscuits. All water entering the body, regardless of source, is of equal value to meeting fluid needs.

People sometimes use thirst as a guide to how much liquids they should drink, but thirst is only the body's first sign of dehydration. By the time we are thirsty, we have lost 2–3 per cent body weight through dehydration. If we lost just 5 per cent of our body's water content, we would be in a state of extreme dehydration. Our mind would become confused and our body very weak (Chatterjee 2018). The symptoms of acute dehydration are headaches, dizziness, nausea, confusion and in some cases seizures and unconsciousness.

When we are dehydrated the blood thickens. Its consistency starts to resemble maple syrup rather than quick flowing apple juice, and our blood cells stop operating adequately. Immune cells cease doing their job effectively because they must inefficiently swim through sludge. Thus, when we experience even moderate dehydration, all membranes become less permeable, hampering the flow of hormones and nutrients into the cell and preventing water products that cause cell damage from flowing out. When this happens throughout our body, our energy is sapped and fatigue can take over. When a person becomes dehydrated, his/her brain shrinks.

Dehydration and Brain Performance

Investigations into dehydration and mental performance were first systematically carried out with a military population (Masento et al. 2014). Soldiers were exposed to extreme heat, including varying degrees of dehydration during service. Cognitive abilities such as short-term memory, numerical ability, psychomotor function and sustained attention were assessed to establish any particular deficits as a result of changes in hydration states. When soldiers were in the severe state of dehydration (below 2 per cent body mass loss) results of cognitive testing showed compromised performance. This study was the first to emphasise that cognitive abilities were sensitive to a suboptimal hydration state.

Subsequent studies both in a military population and in a general population supported this initial evidence of deficits in cognitive ability with induced dehydration. Cognitive deficits can be more modest, depending on the variation in hydration states induced, the target population and age profile of participants. Particular cognitive domains such as short-term memory and perceptual abilities were consistent. Studies measuring self-reported changes in mental states however have consistently found association between dehydration and mood. Mood states most frequently reported used terms such as 'less alert', 'difficulty in concentration', 'fatigue' and 'tension'.

One mechanism for cognitive deficits during dehydration may result from increased cortisol in the blood stream which is released during a stress episode. It has been shown that high levels of cortisol can lower memory function and processing speed and consequently cause memory-related cognitive deficits. Likewise neurotransmitters such as serotonin and dopamine may cause central nervous system dysfunction, influencing brain activity and therefore cognitive performance. Indeed neural activity in brain regions involved in attention and decision-making functions has been shown to decrease when individuals are mildly dehydrated.

Poor hydration affects mental performance and learning ability by reducing the brain's ability to transmit and receive information, while attention and concentration decreases by as much as 10 per cent. All of our cells need water, so our brain is not the only part of our body that is affected by not getting enough of it. Brain cells require a delicate balance between water and various elements to operate, and when we lose too much water the balance is disrupted and brain cells lose efficiency. Having a poorly hydrated brain will make for slower memory, vision problems and even reasoning ability.

Dehydration induced through exercise or heat stress has been associated with decreased short-term memory, long-term memory, arithmetic efficiency, visuospatial function and attention (Cian et al. 2000; Cian et al. 2001; Gopinathan et al. 1988; Baker et al. 2007). Research carried out in the UK also found that drinking water is associated with better attention, short term memory and visual search (Edmonds et al. 2009; Chard et al. 2019; Fadda et al. 2012). Continuous performance tests have been used to assess vigilance-related attentional performance in athletes. For instance, a study by Baker et al. (2007) found that dehydration impaired basketball players' vigilance during games. They found that dehydration slowed response time and contributed to inattention which made players to miss relevant cues. It is also possible that the amount of water drunk may affect performance of tasks differently. Evidence from public surveys have indicated that particular groups are at risk of voluntary dehydration, meaning individuals are drinking insufficient amounts of fluid, resulting in mild sustained dehydration.

Essentially water helps maintain electrolyte (sodium and potassium) balance in the body which is essential for nerve signalling and communication within the brain. It also helps the body to maintain adequate blood volume and blood pressure, both of which are crucial for proper circulation, including blood flow to the brain. Improved blood flow ensures that the brain receives the oxygen and nutrients it needs for optimal functioning. That is why dehydration and poor brain health are closely associated and why increasing our fluid intake can improve our learning capacity, thinking and concentration. When people become dehydrated and perform cognitively engaging tasks, their brains show signs of increased neuronal activation. This means that their brains are working harder than usual to complete such tasks. Increased activity causes fatigue and mood changes in young, healthy adults.

In examinations, drinking water has been shown to have a positive effect on exam performance. (Pawson and Gardner 2013). A study from the University of East London and Westminster found that students that took water into the exam hall scored an average of 5 per cent higher than those without. The researchers observed 447 psychology students, of which 71 were in their foundation year, 225 were first year students and 151 were in their second year. Just 25 per cent of the 447 entered the exam hall with a bottle of water, and it was found that the more mature students were more likely to bring water – 31 per cent of these brought water compared to 21 per cent of foundation year and first year students.

After taking students' academic ability into account, by examining coursework grades, the researchers found foundation year students who drank water during the exam could expect to see grades improved by up to 10 per cent. The improvement was 5 per cent for first year students and 2 per cent for second years. Across the cohort, the improvement in the marks was 4.8 per cent for water-drinking exam candidates. The research findings highlight the importance of staying hydrated during exams and should be targeted at younger students in particular. The findings are also significant because consuming water may have a physiological effect on thinking functions that lead to improved exam performance. Water consumption may also alleviate anxiety, which is known to have a negative effect on exam performance. While further research is needed to tease apart these explanations, it is clear that students should endeavour to stay hydrated with water during exams (Pawson et al. 2013).

Voluntary dehydration is likely to occur due to a lack of awareness of how much fluid consumption is required for a balanced hydration state, especially when not taking into account the amount of daily activity. Other external factors such as social pressures also contribute to this day-to-day variability in hydration requirements. Young adult students are often most at risk due to this transition phase in their lifestyle choices. To ensure that learners have adequate access to water in the USA, *The Healthy, Hunger-Free Kids Act* of 2010 expanded access to drinking water in schools, particularly during meal times. This is supported by the *Centre for Disease Control and Prevention* who advocate that all schools should provide access to water fountains, dispensers and hydration stations throughout the school.

How Much Water Should We Drink?

In recent years, drinking water has become a 'lifestyle' fashion statement. Water 'warriors' can be seen everywhere. They carry bottles of mineral water wherever they go, and they swig and gulp all day, desperate to ensure that they drink their daily 2 litres. Where this idea of '2 litres a day' came from nobody appears to know for certain. According to Smith (2002) this information can be traced back to a 1945 edict from the US Academy of sciences on recommended daily allowances. The last sentence of that edict says that most of the water is 'contained in prepared foods. However, this last sentence has somehow got lost' over time (p. 109).

So how much water should we be drinking? An answer to this question varies depending on what we read and who we listen to. In the USA they have been recommending eight 8-ounce glasses of water every day for many years. According to the popular press and some nutritionists, it is recommended adults should drink 2.5 litres of water daily or 6 to 8 cups or glasses of water, tea, coffee, juice or sugar-free. The European Food Safety Authority (2010) have set recommended guidelines of 2,000 ml of fluids for females and 2,500 ml for males per day. These guidelines were set to encourage more fluid consumption and reduce the risk of sustained dehydration.

There is some debate regarding the guidelines – in particular there is an apparent lack of empirical evidence regarding the volume of fluids an individual should actually consume. Based on the high individual variability in terms of body size and fluid requirements, it is argued that the emphasis should be on encouraging individuals to monitor their own hydration levels using markers such as urine colour and on being aware of variables that influence the amount of water they need to consume (such as physical activity and climate). Most medical professionals agree that making sure we urinate every 2–3 hours and having a light-yellow colour urine means that the body is well hydrated.

Fortunately our bodies can obtain water from multiple nutritional resources and not just from the number of glasses of water we drink per day. Fruits contain the purest water and are loaded with health-promoting minerals. In particular smoothies provide the perfect quick pick-me-up and are a great way to incorporate several servings of fruit and water into our day. When we blend different fruits, we break down their fibres and help release their nutrients. It is these nutrients in the bloodstream that feed the neurons in the brain and strengthen their communication potential. Since many of us do not chew our food thoroughly enough, drinking high-nutrient, raw fruits allow us to digest and assimilate valuable amounts of plant food and water.

We recommend using common sense, avoid dehydration and prioritise water consumption to compensate for any excess sweating. The following tips may help in managing hydration: first thing in the morning drink a glass of tepid water with a slice of lemon in it, keep a water bottle or glass visible as a reminder to drink throughout the day, choose water instead of other beverages during the lunch-time meal, eat fruits and nuts that have high water content as occasional snacks throughout the day.

The Therapeutic Use of Water

Water is the most universal and the most ancient of all remedial agents dating back to the early Egyptian dynasties. In Japan the bath has been used for healing for over a thousand years. Under the Romans the therapeutic use of water reached a peak – a forerunner to the modern hydrotherapy spas of Europe. So how can the therapeutic use of water heal the body and the brain?

When the body is warmed up, either by a hot shower or bath, the blood vessels in the skin expand, increasing their capacity to take more blood. To fill these expanded blood vessels in the skin, blood rushes from the deeper tissues of the body to the skin. Ideally by taking a cool shower immediately afterwards the skin blood vessels contract, causing even more warm blood to go back to the deeper tissues. At the same time the brain records this coolness in the skin area and compensates by sending fresh warm blood from the liver into the circulation. This causes fresh warm blood to go to all parts of the body. This flushing of the deeper tissues of the body with extra fresh warm blood causes a general feeling of warmth throughout the body, even though a cool shower has been taken. It also causes a feeling of vitality and freshness.

Cooling off caused by cold water stimulates the nerves and hormonal pathways to work extra hard to bring back the inner-body temperature to normal (Kenny 2022). The heart rate increases, as does cardiac output, resulting in a sensation similar to the runner's high. A randomised trial showed participants who had cold water stimulus were 54 per cent less likely in the twelve month follow up to take any sick leave for infections and nearly all reported higher energy levels. The duration time can range from 1 to 5 minutes, but it appears we get the same benefit from 30 seconds to a minute. Start with just 5 seconds, Kenny says and build up gradually. A sauna offers similar benefits, as it stimulates the body to counteract the warming effect.

A sauna produces rejuvenation by increasing perspiration. The skin is the body's largest eliminative organ and 30 per cent of all toxins are eliminated by perspiration. In fact chemical analysis shows that perspiration is almost identical to urine. As well as increasing the elimination of toxins, increased body temperature inhibits the growth of viruses and bacteria. A Finnish study involving 2,000 men (between 1984 and 1989) found that moderate to high frequency sauna use was associated with lowered risks of dementia and Alzheimer's disease (Laukkanen et al. 2017: 245).

Conclusion

The therapeutic effects of water on brain health and cognitive ability cannot be over-estimated. 73 per cent of the brain is made up of water and we can only last a few days without it. Drinking water throughout the day prevents us feeling tired, experiencing low-grade headaches, finding it hard to concentrate or comprehending a lot of information. Water helps us digest food and supply glucose and nutrients to feed the brain.

The purpose of this book is to bridge the learning from neuroscience with the practice of learning. The research presented in this chapter provides practical ways that neuroscience can be applied to education. For example, the study by Pawson et al. (2013) demonstrate positive correlation between hydration and cognitive function. These findings clearly show that learners should be mindful to drink sufficient water throughout the day, and particularly during exams. Furthermore, ensuring adequate access to water could be a simple but effective policy response from higher education institutions to support student performance.

References

Amen, Daniel G. (2016). *Change Your Brain Change Your Life*. New York: Piatkus.
Baker, L. B., Conroy, D. E., Kenney, W. L. (2007). Dehydration Impairs Vigilance-related Attention in Male Basketball Players. *Medical Science Sports Exercise*, 39(6), 976–983.
Chard, A. N., Trinies, V., Edmonds, C. J., Sogore, A. and Freeman, M. C. (2019). The Impact of Water Consumption on Hydration and Cognition among Schoolchildren: Methods and Results from a Crossover Trial in Rural Mali. *PLoS One*, 17; 14(1), 1–14.
Chatterjee, R. (2018). *The 4 Pillar Plan: How to Relax, Eat, Move and Sleep*. London: Penguin Random House.
Cian, C., Barraud, P. A., Melin, B. and Raphel, C. (2001). Effects of Fluid Ingestion on Cognitive Function after Heat Stress or Exercise-induced Dehydration. *International Journal of Psychophysiology*, 42(3), 243–251.

Cian, C., Koulmann, N., Barraud, P., Raphel, C., Jimenez, C. and Melin, B. (2000). Influences of Variations in Body Hydration on Cognitive Function: Effect of Hyperhydration, Heat Stress, and Exercise-induced Dehydration. *Journal of Psychophysiology*, 14(1), 29.

Edmonds, C. J. and Burford, D. (2009). Should Children Drink More Water? The Effects of Drinking Water on Cognition in Children. *Appetite*, 52, 776–779.

Edmonds, C. J. and Jeffes, B. (2009). Does Having a Drink Help You Think? 6–7 Year-Old Children Show Improvements in Cognitive Performance from Baseline to Test after Having a Drink of Water. *Appetite*, 53, 469–472.

European Food Safety Authority (EFSA). (2010). Scientific Opinion on Dietary Values of Water. *EFSA Panel on Dietetic Products*, J8, 1459.

Fadda, R., Rapinett, G., Grathwohl, D., Parisi, M., Fanari, R. and Calo, C. M. (2012). Effects of Drinking Supplementary Water at School on Cognitive Performance in Children. *Appetite*, 59(3), 730–737.

Gopinathan, P. M., Pichan, G. and Sharma, V. M. (1988). Role of Dehydration in Heat Stress-induced Variations in Mental Performance. *Arch Environ Health*, 43(1), 15–17.

Kempton, M. J., Ettinger, U., Foster, R., Williams, S. C. R., Calvert, G. A. and Hampshire, A. (2011). Dehydration Affects Brain Structure and Function in Healthy Adolescents. *Human Brain Mapping*, 32, 71–79.

Kenney, E. L., Long, M. W., Cradock, A. L. and Gortmaker, S. L. (2015). Prevalence of Inadequate Hydration among US Children and Disparities by Gender and Race/Ethnicity: National Health and Nutrition Examination Survey, 2009–2012. *American Journal of Public Health*, 105(8).

Kenny, A. (2022). *Age Proof: The New Science of Living a Longer and Healthier Life*. London: Bonnier Books.

Laukkanen, T., Kunutsor, S., Kauhanen, J. and Laukkanen, J. A. (2017). Sauna Bathing Is Inversely Associated with Dementia and Alzheimer's Disease in Middle-aged Finnish Men. *Age Ageing*, 46(2), 245–249.

Mahes, P. (n.d.). in Anderson, B. (2020). *Prayers for Calm: Mediations Affirmations and Prayers to Soothe Your Soul*. San Francisco: Mango Coral Gabels.

Masento, N. A., Golightly, M., Field, D. T., Butler, L. T. and van Reekum, C. M. (2014). Effects of Hydration Status on Cognitive Performance and Mood. *British Journal of Nutrition*, 111(10):1841–1852.

Pawson, C., Gardner, M., Doherty, S., Martin, L., Soares, R. and Edmonds, C. (2013). Drink Availability Is Associated with Enhanced Examination Performance in Adults. *Psychology Teaching Review*, 19(1), 57–66.

Popkin, B. M., D'Anci, K. E. and Rosenberg, I. H. (2010). Water, Hydration, and Health. *Nutrition Reviews*, 68(8), 439–458.

Smith, A. (2002). *The Brain's Behind It: New Knowledge about the Brain and Learning.* Stafford, UK: Network Educational Press.

UNICEF (2012). *UNICEF Water, Sanitation and Hygiene Annual Report 2012.* New York: UNICEF WASH Section, Programme Division; https://www.unicef.org/media/92866/file/UNICEF-annual-report-2012.pdf.

Valtin, H. (2002). Drinking at least Eight Glasses of Water a Day. Really? Is Their Scientific Evidence for 8×8? *American Journal of Physiology. Regulatory, Integrative and Comparative Physiology*, 283: R993–R1004.

CHAPTER 16

Empathy for Brain Health

> 'Empathy is seeing with the eyes of another, listening with the ears of another, and feeling with the heart of another'.
>
> (Adler 1956)

Empathy seems to have deep roots in the human brain and body, and in our evolutionary history. Elementary forms of empathy have been observed in our primate relatives as well as in dogs and rats. Empathy has been associated with two different pathways in the brain, and scientists have speculated that some aspects of empathy can be traced to mirror neurons – cells in the brain that fire when we observe someone else perform an action in much the same way that they would fire if we performed the action ourselves.

The American Psychology Association defines empathy as 'understanding a person from their frame of reference rather than one's own or vicariously experiencing that person's feelings, perceptions and thoughts'. It refers to the ability to share someone else's feelings or experiences, coupled with an ability to imagine what it would be like to be in that person's situation. It is what Atticus Finch is talking about in the novel *To Kill a Mockingbird* when he says that we never really understand someone until we 'climb into his skin and walk around in it' (Lee 1960).

Research has uncovered evidence of a genetic basis to empathy, though studies suggest that practice can enhance (or restrict) a person's natural empathic abilities. It is often differentiated into two types: affective and cognitive empathy (Penton 2017) Affective empathy refers to the sensation and feelings we get in response to other's emotions, while cognitive empathy refers to our ability to identify and understand other people's emotions. Empathy is an important component of emotional intelligence and essential for building strong relationships with others.

As far as the brain is concerned, empathy is both a trait and a skill. We internalise the way our caregivers respond to stress and the way they are or are not able to depend on others for help and support. We tend to grow up to care about other people and ourselves the same way we were first cared for. Of course, this depends on many factors but the nature of the initial nurturing/ bonding relationship is powerfully influential. One of the most exciting realities about brain health is that since the brain changes through repeated experiences, there are plenty of ways we can cultivate empathy in the everyday interactions of our family, friends, colleagues, neighbours and the wider community. In this chapter we will explore the many ways empathy may be expressed or experienced, how it can be learned and the positive role it can play in brain health and learning.

The Self and the Other

The subtlety and diversity required of the brain to perform the complex task of perceiving and relating to another's mental state is reflected in the range of brain regions associated with such a task. This is known as the empathy loop. The main area involves the ventromedial (front and middle) part of the frontal cortex. This part of the brain serves as a region for binding together the large-scale networks that are reciprocally interconnected with the amygdala, home to emotional processing, self-regulation, recognition making, self-perception and empathy.

Empathy is neither agreement nor approval, but in the words of Bullard the 'highest form of knowledge' (Bullard 2012). Empathy is vital in building successful interpersonal relationships of all types, in the family unit, workplace and beyond. Identity becomes stronger when we define our 'selves' in relation to an 'other' (Foucault 1994). We can empathise with someone we wish would act differently. Empathy doesn't mean waiving our rights or having low standards of expectations. Knowing this can help us feel it is okay to be empathic.

Researchers on burn-out in human service organisations also talk about unhealthy empathy, which can occur when a person over-identifies with someone else's feelings, leading to them feeling overwhelmed and so immersed in the other person's problems they are unable to deal with their own. In the literature this is sometimes called the burn-out syndrome (Kahn et al. 1994). When the task of empathy is perceived in this way, a person strives to make everybody feel better. Constantly trying to please others and seeking to come to everyone's emotional rescue can lead to stress and anxiety.

Research reveals that when we reach out to help others rather than over-identify with their hurt as our own, we must retain a sense of equilibrium. Over-identifying can be emotionally taxing, especially for individuals who are repeatedly encountering distressing situations. Empathising can therefore be emotionally exhausting and can lead to feelings of depletion. Tibetan Buddhist monks recognise and practice compassion rather than empathy as a way to protect against the potential fatigue and emotional detachment stemming from empathy (Davidson and Harrington, 2002).

Contemporary researchers distinguish between three types of empathy: empathy as an emotional response, as a cognitive response and as a compassionate response. However each manifest themselves in different ways and it is probably more helpful to understand them as three different levels, involving many separate neural pathways and network loops in the brain (Bazalgette 2017).

Emotional Empathy

Emotional (affective) empathy refers to the sensations and feelings we get in response to others' emotions. It is the ability to respond to other people's emotions with sensitivity. It includes mirroring the actions, sensations and feelings of another as we try to imagine what it is like to be in their situation and what it must feel like. We want to share their distress and, at the same time, attempt to communicate the powerful healing message that we are not alone.

Emotional empathy is self-focused in that when we sense another's pain or stress our own repertoire of emotional memories of a similar event are triggered. In other words, we first experience our own feelings before we can relate to the situation and emotion of the other. In a sense emotional empathy is only the first step in understanding the actions of others, but plays a crucial role helping us to understand 'from the inside', actions we already know how to perform. It starts in a part of the brain first developed to enable us to detect what was dangerous and what was safe. It is a key survival tool and one we share with many animals.

Take for example a wild reindeer herd grazing peacefully in Lapland. One of them notices a wolf. That reindeer's emotional state of arousal and alarm spreads throughout the herd like lightning and, as if with one mind, they scatter to safety. There is no 'thought' involved. This is because most animals have the same basic nervous system and parts of a brain that can detect things like changes in posture, behaviour and movement.

Emotional empathy is actually deeply rooted in a human's brain through mirror neurons and neuroscientists use this to explain emotional empathy. How we read faces so rapidly and automatically is known as mirroring, based on the simultaneous workings of mirror neurons in the brain. For example, someone we know comes to us in tears, so it is natural to feel that same pull on our own heartstrings. There are muscles everywhere under the skin (except for the tip of the nose and the chin), and so facial expressions are made up of muscle movements. The theory of mirror neurons is that people are continually supported by the automatic mimicry of mirror neurons to understand the emotions of others, and what is happening around them.

It works something like this. Our facial muscles reflect what we are feeling, and our neural machinery takes advantage of that. When we are trying to understand what we are feeling, we try on our facial expression. We do not mean to do so – it happens rapidly and unconsciously – but that automatic mirroring of our expression gives us a rapid estimate of what we are likely to be feeling. This is a powerful shortcut for our brain to gain a better understanding of our self and make better predictions about what we say and do.

The brain is specifically adapted to social openness, and mirror neurons fire when an individual imitates the actions of those who are tuned to register other people's emotions and intentions. For every moment of our lives, our brain circuitry decodes the emotions of others based on extremely subtle facial cues. We learn this very early in childhood as we instinctively know when our caregiver's face changes, or when their body movement becomes more rigid and there is a particular tone to their voice.

Cognitive Empathy

Imagine the following – you are in a good mood when your partner comes home. However you notice that your partner's face is tight, and their belongings are plonked a little too firmly on the floor. You sense something is 'not right' and you immediately feel concern or even alarm in your own body. You are not just feeling your partner's distress, but you also want to figure out what you might do to alleviate some of it. You might soften your face and ask empathically about what's up and try to appraise your partner's experience or to try to understand why they are feeling the way they do. This is called cognitive empathy. You are not just reacting at an emotional level, but you are using cognitive thought processes to understand their mental state.

Cognitive empathy is a deliberate process of imagining what another person might be thinking. It is also known as perspective taking because it is 'other focused' and concerned with knowing, understanding or comprehending on an intellectual level. Clearly understanding anger is not the same thing as feeling angry. Perspective taking is a cognitively demanding task that generally involves deliberate effort.

Many people believe that cognitive empathy is more like a skill rather than a trait as it is helpful in negotiations, motivating other people and understanding diverse viewpoints. It is considered an important component of emotional intelligence according to Daniel Goleman. It is the entirely human ability to put oneself figuratively speaking into the shoes of another 'brain'. It is a social skill that requires the ability to understand, as far as possible, the needs and intentions of another person. This ability is sometimes called theory of mind.

This theory of mind tends to be used with regard to children. That is a child with a theory of mind is a child who recognises that people have mental lives, beliefs, desires and dreams. This ability gives the child a framework to explain the behaviour of others. While children as young as 18 months to 2 years may have a rudimentary theory of mind, the general consensus is that it is only fully formed between 3 and 4 years of age and linked to the development of speech and language.

The classic way to measure a theory of mind is the 'false-belief test' (2016). This is a test used with young children to assess their understanding of reality and how other people perceive it. Children are asked to predict how some deceptive object will appear to another person. For example, children are shown a chocolate box that turns out to hold pencils. They are asked what someone else will expect to see when they open the box. Children around the age of three consistently believe the other person will expect to see pencils, older children correctly believe that others will expect chocolates. In this task, a child is shown and asked to identify objects that appear to be one thing but are actually something else.

The roots of empathy start early in childhood and while we are born hardwired with the capacity for empathy, its development requires experience and practice. As we have seen empathy is both an emotional and cognitive experience. The emotional components are the first to emerge. Babies begin reflecting the emotional states and experiences of those around them right away while older children who feel safe, secure and loved are eventually more sensitive to others' emotional needs. As children get older, the cognitive component gradually complements the emotional template. By pre-school age, children become more aware that other people have separate bodies, feelings and experiences. They develop a theory of mind which enables them to engage in early 'perspective taking', a precursor to being able to empathise. For example, a 1-year-old sees a friend is upset, so he may go to his own mother to comfort him. A 3-year-old in the same situation may go and get his friend's mother because he now understands that his friend would want his own parent in a time of distress. The distinction between self and others matures quickly throughout early childhood.

This ability to introspectively perceive one's own mind, and to understand another person's mental state, gives us the unique ability to bond together, communicate with each other, imagine new possibilities and adapt creatively to changing circumstance.

Compassionate Empathy

Compassion is to feel sorrow or pity for the suffering of another and is associated with helping or altruistic behaviour. This behaviour, as understood in its pure form, is motivated by the desire to benefit another with no expectation of personal gain or reward. Such kindness depends on the prefrontal brain stimulating intentions, limbic-based emotions and rewards, as well as the neurochemical oxytocin and brain stem arousal. When we are kind to someone else, we also benefit our self as it feels good to be kind and it encourages others to treat us well in turn. As Mackesy (2019) writes, 'nothing beats kindness, it sits quietly beyond all things'.

There is strong evidence to suggest that both children and adults respond empathically to the distress of another person, suggesting that it is unpleasant to watch someone else suffer. Even infants as young as a few months can respond to the distress of another infant. This would seem to imply that there is a biological mechanism that may predispose a person with the urge to react but how a person responds is also influenced by past experience and immediate circumstances.

Many of the most inspirational examples of human behaviour, including aiding strangers and stigmatised people, are thought to have empathic altruistic roots. We have all heard anecdotes of people going to extraordinary lengths – entering burning buildings, stepping in front of moving cars or jumping into rivers – to save someone they did not know. However deciding to help or not involves a complex brain process and depends on how we interpret the event and consider the personal benefits and costs of not helping.

The behaviour of others around us can influence what we do. As noted above a major prediction about an emergency is that a particular individual will respond quiet differently according to whether others are present or not. The reason for this is what researchers call 'diffusion of responsibility', or the 'bystander intervention' effect (Beyer et al. 2017). Namely an individual who is part of a group often tends to offload responsibility for action onto others. In the case of an emergency situation, the presence of other onlookers provides the opportunity to transfer the responsibility for action, or inaction, onto them. People who are alone are most likely to help a victim because they believe they carry the entire responsibility for action. Furthermore, when bystanders believe they are similar to the victim, and identify with that person, they experience empathic concern.

Consider the questions that might run through our brains should we encounter a possible emergency as we walk down a city street. At first our natural tendency to experience unpleasant feelings at the sight of another person in pain is triggered in the brain. The more the victim is in pain, the greater is our unpleasant feeling. We can reduce this unpleasant feeling by helping the victim or psychologically removing ourselves from the situation. If, for instance we are with a friend or partner we are much more likely to respond. Empathy is vital in building successful interpersonal relationships of all types, in the family unit, workplace and beyond. However from a mental health perspective, those who have high levels of empathy are likely to function well in society, reporting larger social circles and more satisfying relationships.

Learning and Practising Empathy

Empathy is considered by many as one of most important skills for the twenty-first century. In this context the question is whether skills like empathy can be trained and improved or is it just another chance trait that some lucky people are born with? Over the last few years scientists have delved much more deeply into empathy, and they have discovered that while we are born hardwired with the capacity for empathy its development requires experience and practice.

The science of neuroplasticity assures us that all experiences shape the brain. This insight tells us that since the brain changes through repeated experiences, there are plenty of ways to cultivate empathy. In their book *The Yes Brain* Siegel and Bryson firmly believe that empathy can and should be taught to children in school. They called this strategy 'fine-tuning children's empathy radar' (2018: 40). We believe this concept equally applies to learners of all ages. We use the metaphor of 'empathy radar' as a brain health strategy to mindfully activate the brain's social engagement system, in such a way that we view situations through the lens of empathy and caring for others. An active 'empathy radar' helps us notice both the verbal and nonverbal signals of what may be going on in other people's mind. When consciously activated we become more mindful and more receptive to understanding another person's state of mind. It might mean simply being more aware of times when we have not listened attentively to someone or when we may have dominated a conversation.

Kabat-Zinn (1994) emphasises the importance of recognising that we are more than the thinking part of our brain, saying 'when we look at thoughts as just thoughts, purposefully not reacting to their content and to their emotional charge, we become at least a little freer from their attraction or repulsion'. The busy brain is only one aspect of who we are, but we tend to live most of our mental time within the clutter of past or future events. This can be counteracted through creating a conscious intention to live each moment fully and thus more open to engaging our empathy radar. For instance, Bergland (2007) argues that compassion can be taught and learned through rigorous practice that includes loving kindness meditation (LKM) which can rewire our brain. Practising LKM is easy. All we have to do is take a few minutes every day to sit quietly and systematically send loving and compassionate thoughts to our self and others. Doing this simple four-step LKM as practice rewires our brain by engaging neural connections to empathy.

Bazalgette (2017) proposed some key ideas which include further mapping of the brain's empathy circuits. He suggests that we must ensure that every child gets the one-to-one nurture and stimulation they need to give them their own functioning empathy circuits. Educationalists can assess and cultivate the emotional intelligence of every learner, and policymakers can ensure that the arts and popular culture are promoted to that help us understand the perspectives of others.

Conclusion

Empathy is an important attribute for individuals but it also contributes to social cohesion. An important facet of empathy is cultivating empathy for one's self. For learners having empathy for others can help them to emotionally connect with and demonstrate respect for alternative perspectives. Empathy for one self can help learners to better understand their own emotions and intentions, to improve their communication and to connect more meaningfully with the material they are learning. Teaching with empathy can help a teacher create a more supportive and inclusive environment. In this chapter we have described how, through MRI scanning technology, neuroscientists have recognised the importance of empathy for learning and for brain development.

References

Ansbacher, H. L. and Ansbacher, R. R. (Eds). (1956). *The Individual Psychology of Alfred Adler*. Basic Books.

Bazalgette, P. (2017). *The Empathy Instinct: How to Create a More Civil Society*. London: John Murray Publishers.

Bergland, C. (2007). *The Athlete's Way: Sweat and the Biology of Bliss*. London: St Martin's Press.

Beyer, F., Sidarus, N., Bonicalzi, S. and Haggard, P. (2017). Beyond Self-serving Bias: Diffusion of Responsibility Reduces Sense of Agency and Outcome Monitoring. *Social Cognitive and Affective Neuroscience*, 12(1), 138–145.

Blakeslee, S. and Blakeslee, M. (2007). *The Body Has a Mind of Its Own*. New York: Random House.

Bullard, B. (2012). *Tight Turns: Memoirs, 20th Century, Coal Fields, World War II*. Create Space, Independent Publishing Platform.

Calvo-Merino, B., Glaser, D. E., Grezes, J., Passingham, R. E. and Haggard, P. (2005). Action Observation and Acquired Motor Skills: An FMRI Study with Expert Dancers. *Cerebral Cortex,* 15(8), 1243–1249.

Davidson, R. J. and Harrington, A. (Eds). (2002). *Visions of Compassion: Western Scientists and Tibetan Buddhists Examine Human Nature.* Oxford: Oxford University Press.

Foucault, M. (1994). *The Order of Things: An Archaeology of the Human Sciences.* A translation of Les Mots et les choses. New York: Vintage Books.

Goleman, D. (1998). *Emotional Intelligence: Why It Can Matter More Than I.Q.* New York: Bloomsbury Publishing.

Kabat-Zinn, J. (1994). *Wherever You Go There You Are.* New York: Hyperion.

Kahn, R. et al. (1994). *Organisational Stress: Studies in Role Conflict and Ambiguity.* New York: John Wiley and Sons.

Lee, H. (1960). *To Kill a Mockingbird.* Philadelphia: J. B. Lippincott and Co.

Mackesy, C. (2019). *The Boy, the Mole, the Fox and the Horse.* London: Ebury Publishers.

Macknik, S. and Macknik M. (2011). *Sleights of Mind: What the Neuroscience of Magic Reveals about Our Brains.* London: Profile Books.

Peyton, S. (2017). *Your Resonant Self: Guided Meditations and Exercises to Engage Your Brain's Capacity for Healing.* New York: W. W. Norton and Company.

See, A. (2016). The False-Belief Test. In V. Weekes-Shackelford, T. Shackelford and V. Weekes-Shackelford (Eds), *Encyclopedia of Evolutionary Psychological Science.* Cham: Springer.

Siegel, D. and Bryson, T. (2018). *The Yes Brain: How to Cultivate Courage, Curiosity, and Resilience in Your Child.* New York: Bantam Books.

Glossary

A.

Action potential: a shift in the neuron's potential electric energy caused by the flow of charged particles in and out of the membrane of the neuron.

Adaptive intervention: a means of communicating without words. A technique used to address the communication gap when people with dementia can no longer use words.

Adrenaline: a hormone that stimulates glucose release as a way of dealing with short-term stress.

Alzheimer's disease: brain disease that slowly destroys brain cells and brain cell connections (synapses) by progressive loss of memory, concentration and judgement and by personality changes. Eventually, victims are unable to recognise other people and to control body functions.

Amygdala: a pair of almond shaped structures in the brain involved in processing emotions and forming memories. This is the primitive emotional brain centre responsible for detecting the level of stress and fear in your environment. It releases the stress hormone cortisol and adrenaline, which help prepare the body to stand and fight or run away in flight, otherwise known as the 'fight-or-flight mechanism'.

Anxiety: a state in which a person experiences stress, worry, nervousness or unease. The nerves we feel before making a speech or prior to a sporting competitive event are examples of anxiety. In this book anxiety was discussed because it is highly correlated with gut disorders, and often healing the gut can rid us of certain anxiety behaviours.

Atrophy: as neurons are injured and die throughout the brain, connections between networks may break down, and many brain regions begin to shrink. By the final stages of Alzheimer's disease, this process is called atrophy, resulting in significant loss of brain volume.

Autopilot mode: a habit of mind that kicks in when we allow what was initially novel and which has now become mundane or routine.

Axon: a long, threadlike nerve fibre that shuttles (transports) electrical impulses away from the main body of a neuron.

Active Learning: an instructional approach that engages students in the learning process through activities and experiences that require them to actively participate and interact with the material. Active learning is designed to promote deeper understanding, critical thinking and the application of knowledge.

B.

Biomarker (biological marker): a characteristic that can be measured to indicate the state of health or disease of an individual.

Blood-brain barrier: the mechanisms that keep certain toxic substances in the blood from penetrating brain neural tissue.

Brain Health: is continually challenging the brain with new activities by promoting the growth and maintenance of new and existing neurons. Optimal brain health involves the continual stimulation of the neurons for maximum efficiency.

Brain integration: describes what happens when differentiated parts of the brain work together.

Brainstem: a primitive part of the brain that is responsible for many basic functions, such as heartrate and breathing. It also regulates the central nervous system and the sleep cycle. It is the major route by which the forebrain communicates with the spinal cord and peripheral nerves.

Behaviourism: is a psychological theory and approach that focuses on observable behaviours and external stimuli, emphasising the role of the environment in shaping and controlling behaviour associated with Pavlov and Skinner.

Bloom's Taxonomy: considered one of the most useful tools for moving students to higher levels of thinking ranging from understanding, comprehension, application, to higher order evaluation.

C.

Cerebrospinal Fluid: liquid that surrounds the brain and spinal cord; it plays a role in cushioning those structures as well as the transport of substances between the brain tissue and bloodstream.

Cerebrospinal Disease: disease of the blood vessels that supply the brain, causing problems with blood flow to affected regions.

Circadian Rhythm: a daily 24-hour sleep-wake cycle is considered to be a circadian rhythm.

Cognition: a broad psychological term used to refer to such activities as thinking, conceiving and reasoning.

Cognitive Load: the total amount of mental effort or resources required to perform a particular cognitive task. Cognitive load theory, developed by John Sweller, focuses on the limitations of working memory and how instructional design can be optimised to manage cognitive load for more effective learning.

Cognitive Development: children go through specific stages of learning in response to the way that they adapt to their environment.

Cognitive Reserve: suggests that keeping cognitively stimulated throughout life may play a role in building resilience to help with early disease processes. Engaging in stimulating brain activity may assist in creating increased cognitive reserve, enhancing brain adaptability and compensating for other damaged areas. The brain is a muscle that requires exercise as do other body muscles.

Cognitive Reserve Theory: the more we use our brains the more able our brains are to cope with the effects of dementia.

Cognitive Test: a series of set tasks designed to assess thinking and memory ability.

Consolidation: in the context of learning and memory, refers to the process by which newly acquired information is stabilised and strengthened, making it more resistant to forgetting.

Constructivism: a learning theory and educational philosophy that emphasises the active role of learners in constructing their understanding and knowledge of the world. It suggests that learning is a process of mental construction where individuals build new knowledge and meaning based on their existing cognitive structures, experiences, and interactions with the environment.

Corpus Callosum: the intricate bundle of fibres that connects and enables communication between the two hemispheres in the brain.

Cortex: the thin, highly folded outer layer of the brain. It coordinates some of our most advanced mental functions, like planning, language and complex thoughts.

Creativity: commonly understood as the ability to imagine and create things in new ways.

D.

Dopamine: a neurotransmitter that is used in our brain and contributes to the reward system, which can make us feel good, and possibly underlies addiction.

Dementia: umbrella term for a group of brain symptoms associated with impairments in memory, thinking or behaviour which impact the ability to perform everyday activities independently.

Dendrite: a short, slender extension from the cell body of a neuron.

Discovery Learning: an instructional approach that emphasises student-centred and inquiry-based learning, where learners actively discover and construct knowledge through exploration, problem-solving, and direct experience.

Dual Coding: a cognitive learning theory that suggests that people can process and retain information more effectively when it is presented in both verbal and visual formats. The theory, developed by cognitive psychologist Allan Paivio, proposes that there are two cognitive subsystems involved in human information processing: one for verbal information and one for non-verbal or visual information.

E.

Educational Neuroscience: an interdisciplinary field that brings together principles and methods from neuroscience, psychology, and education to investigate the neural mechanisms underlying learning and educational processes.

Ego: that part of us that is always in search of power, control and approval. There will never be enough to satisfy the ego, which by nature is independent and insatiable. The driving force of the ego is fear of losing what one has.

Endorphins: a category of chemicals produced by the body that have effects similar to those of opiate drugs.

Enteric Nervous System: an interconnected system of neurons that control the gastrointestinal system – is often referred to as our second brain.

Empathy: the perspective that allows us to keep in mind that each of us is not only a 'me' but part of an interconnected 'we' as well.

Enriched Environment: an environment in which the brain is allowed to thrive through the presence of a variety of stimulations and activities.

Executive Abilities: a catch-all term for a wide range of planning, emotional regulation, and goal directed activity skill associated with the frontal lobe.

Executive Control: the ability to filter out distractions, keep hold of a train of thought and focus on a task.

Engagement: engaged individuals actively participate, contribute and invest effort in the activity rather than being passive or disinterested. Cognitive engagement involves mental effort and concentration.

Extrinsic Motivation: often associated with tangible rewards such as money, prizes, grades, or other forms of recognition. Individuals are motivated to perform a task because they expect to receive something external in return.

F.

FMRI: meaning Functional Magnetic Resonance Imaging measures blood flow in the brain to show which parts of the brain are active when a person carries out a particular task.

Faculty Psychology: this underpins notions of general powers of the brain and the existence of general thinking skills such as observation, judgement, imagination and critical thinking.

G.

Gamma-aminobutyric acid (GABA): neurotransmitter that calms the body's stress response.

Genes: units of instruction for heritable traits or those tiny units of hereditary material (DNA) that carry instructions for forming all the cells in the body directing their activity and for behaviour and emotions (Greek word for 'to be born').

Glial Cell: a type of brain cell that provides a supportive role to neurons, repairing damage and so on. Glial cells do not transmit electrical signals but they are vital to the healthy working of the brain.

Grey Matter: the dendrites of neurons appear as 'grey' under brain imaging techniques.

H.

Habituation: is the term for our brain's ability to decrease its response to continuing stimuli.

Hippocampus: so called because it is a seahorse-shaped part of the brain that is involved with memory and emotion. It needs to be well-balanced (stimulated and rested in equal measure) to develop correctly.

Homeostasis: a state of balance between the brain and the workings of the body. The brain receives information from the external world through the senses and from the internal world through the workings of the body.

I.

Insomnia: sleep disorder characterised by difficulty falling or staying asleep, frequent waking during the night, or waking too early.

Interleaved Learning: a learning strategy that involves mixing or alternating different topics or types of problems within a single study session. This is in contrast to 'blocked' or 'massed' learning, where individuals focus on one type of task or topic for an extended period before moving on to another.

Intrinsic Motivation: refers to engaging in an activity or behaviour because it is inherently rewarding and enjoyable, without the need for external rewards or incentives.

Information Processing Theory: a cognitive framework that views the mind as a computer-like system for processing information. This theory emerged in the mid-twentieth century and draws on the analogy between the human mind and a computer to explain how individuals encode, store, retrieve and manipulate information.

K.

Kaizen: a Japanese word used to mean gradual improvement by making small incremental steps. A gradualist approach to goal setting.

L.

Learned Helplessness Theory: refers to the emotional state experienced when one perceives a lack of control over one's situation or environment. Consequently, individuals resign themselves to this negative view and accept their supposed ineffectiveness.

Learning: in common parlance the word 'learning' carries at least two meanings. There is a general one implying some kind of change, often in knowledge. There is also a more intense sense of the verb 'to learn' meaning to memorise, to learn by heart, and to comprehend.

Long-term Memory: involves those memories held and kept available for retrieval at a later date, which include information gleaned over the lifespan. There are three components of long-term memory: (1) memory for events and experiences is known as episodic (2) memory for facts and concepts is known as semantic memory and (3) procedural memory includes memory for skills and tasks.

Lumbar Puncture: medical investigation when a hollow needle is inserted into the lower back to collect a sample of cerebrospinal fluid.

M.

MRI: stands for Magnetic Resonance Imaging. MRI scanners use a strong magnetic field and radio waves to build a 3D picture of a body part and its internal structure.

Melatonin: a hormone secreted by the pineal gland during the hours of darkness which induces sleep.

Metacognition: refers to the ability to think about and regulate one's own cognitive processes. It involves awareness, monitoring, and control of one's thoughts, learning strategies and problem-solving skills.

Motivation: refers to the internal or external factors that drive an individual to initiate, sustain and direct their behaviour towards the accomplishment of specific goals or outcomes.

Mild Cognitive Impairment (MCI): disorder of thinking and memory where a person has problems over and above those expected for their age but these problems do not interfere with their day-to-day living. MCI can, though does not always, progress to dementia.

Myelin: acts as an insulator, and helps neural signals travel more quickly and over greater distances than they could in an uninsulated axon.

Myelin Sheath: white protective covering (sheath) that insulates nerve fibres or covering the brain cell acts like insulation covering electrical wires. It speeds transmission of electrical signals to the brain. Deficiencies in nutrition delay nerve-impulse transmission, negatively impacting brain function.

Myelination: formation of the *myelin* sheath that becomes stronger and life-lasting mainly during adolescence.

N.

Nature-nurture issue: the debate about the relative contributions of nature (genes) and nurture (experience) to the behavioural capacities of individuals.

Neuron: a nerve cell that sends and receives signals across the brain. These are the basic cells of our brain that are in charge of computing, memory and messaging. They have dendrites like tree branches coming off of their cell body, and typically have a long axon that communicates a nerve impulse over distances.

Neurodegenerative Disease: progressive disease of the Central Nervous System that is caused by degradation, dysfunction and death of neurons.

Neurogenesis: the term used by neuroscientists to describe the production of new neurons.

Neuroscience: the scientific study of the nervous system.

Neurotransmitters: chemical messenger molecules released by neurons at the synapses to pass on messages from one neuron to another such as serotonin, dopamine and norepinephrine, and affect mood as well as thoughts and actions.

Non-rapid Eye Movement (NREM) sleep: the first three stages of sleep increasing in depth as they progress.

Neuromyths: are misconceptions or misinterpretations of neuroscience-related information that circulate widely, often in educational or popular science contexts. These myths may arise from a misunderstanding or oversimplification of complex neuroscience concepts, and they can influence beliefs and practices in education.

Neuroimaging: a set of techniques and technologies used to create detailed images of the structure, function, and activity of the brain. Neuroimaging plays a crucial role in research, diagnosis, and treatment planning for various neurological and psychiatric conditions, including Alzheimer's disease, epilepsy, schizophrenia and traumatic brain injuries.

O.

Occipital Lobe: located at the back of the brain. Mostly devoted to vision. Contains the primary visual cortex.

Oxytocin: known as the love hormone because it is associated with physical bonding or helping other people (e.g. volunteering is good for brain health because it releases oxytocin). Touch can be soothing and pleasurable and is often used to convey positive signals such as sympathy, reassurance and comfort. Evidence for the positive effects of touch has been linked to the release of oxytocin, a neuropeptide that is part of the stress response cycle.

P.

Parkinson's Disease: a disease caused by gradual destruction of the substantia nigra in the brain, an area normally rich in dopamine, resulting in loss of voluntary movement.

Pituitary Gland: a small grand at the base of the brain that secretes hormones that regulate most of the other glands in the body. Often referred to as the 'master gland'.

Placebo Effect: from the Latin 'I will please', is the finding that people will feel better from a drug or medical treatment if they believe that it will do them good, even if that 'drug' is just a 'sugar' pill. In research trials is refers to a substance that looks exactly like a drug being tested, so that the patient and the doctor may not know which is which.

Plasticity: the neuroscientific term used to refer to the dynamic, constantly reorganising, and malleable properties of the human brain in response to new experiences or change. In other words, the brain can change itself with new experiences.

Prefrontal Cortex (PFC): the most forward part of the brain, just behind and above the forehead, and our 'cortex' is the outer layer of the brain (its Latin root means 'bark', like the bark of a tree).

Pruning: the brain has the capacity to learn and absorb much information, but the areas of the brain and the type of processes that are most utilised will be retained and strengthened, while those that are not used will fall away. This is illustrated most clearly with language acquisition; although all children are initially able to learn all languages, the ability to form certain sounds, if not utilised, will be more difficult or impossible to pick up later in life.

Problem-based Learning: an instructional approach that focuses on students working collaboratively to solve complex, real-world problems by applying critical thinking, problem-solving skills and subject knowledge to develop a solution.

Patterning: in cognitive and behavioural sciences, patterning refers to the process of recognising and learning patterns of behaviour or thought. This can be relevant in the study of cognitive development, language acquisition, and the formation of habits.

R.

Reticular Activation System (RAS): a collection of nerves in the brainstem that act as a filter for all of the thousands of sights and sounds that our brains perceive on a moment-to-moment basis.

Rejection: to be socially excluded from others has a major negative impact on a person's existence. The impact of social rejection is regarded to be so significant that it has even been related to physical pain and is thought to be controlled by the same physiological system.

REM Sleep: rapid eye movement sleep is a phase of sleep in which our muscles relax and we have vivid dreams. REM sleep is thought to be important for consolidating memories.

Resilience: a state of resourcefulness that lets us move through challenges with strength and clarity. The ability to bounce back from adversity.

Retrieval Practice: a learning strategy that involves actively recalling information from memory as a means of strengthening and enhancing long-term retention.

Reinforcement: in the context of psychology and learning theory, refers to a process by which a stimulus or event strengthens or increases the likelihood of a specific behaviour occurring in the future. Positive Reinforcement involves the presentation of a rewarding stimulus or event following a behaviour, with the aim of increasing the likelihood that the behaviour will be repeated.

S.

Serotonin: the oldest neurotransmitter in the brain, important for emotional processing and sleep.

Sleep Efficiency: The difference between how long we spend in bed versus how long we sleep.

Stress Response: also known as 'fight or flight' response. The sequence of physiological changes in the body in response to a perceived threat.

Stress: the body's response to the brain perceiving a threat.

Subcortical: literally located beneath the cortex, referring to brain structures that are not a part of the cerebral cortex.

Synapses: a small gap between two connecting neurons where signals are passed from one neuron to another to spread information. The junctions of communication between two neurons.

Synaptic Transmitters: a chemical released by a neuron at a synapse that affects the activity of a second neuron.

Scaffolding: in the context of education and learning refers to a teaching method and support system that provides temporary assistance to learners to help them acquire new knowledge, skills, or concepts.

Socio-cultural Learning: refers to the process of acquiring knowledge, skills, values and behaviours within a social and cultural context. This perspective on learning emphasises the significant role that social interactions, cultural influences, and shared experiences play in shaping an individual's understanding of the world and their abilities.

Spaced Learning: a learning strategy that involves breaking up learning material into shorter, spaced intervals over time rather than studying it all at once in a single session.

Self-Regulation: refers to the ability to manage one's thoughts, emotions, behaviours, and impulses in a way that aligns with long-term goals, values and social expectations.

T.

Testosterone: the hormone that defines 'maleness'. It has been linked with male aggressiveness, male libido and impatience.

V.

Validation Therapy: is based on recognising and validating the individual's subjective reality. It is primarily used to provide caregivers with a means of communicating with older people who are experiencing dementia.

Vagus Nerve: a nerve that wanders from the brain through your torso, sending projections to all your organs. It is an important component of the gut-brain-axis, connecting your intestines to your brain.

Vascular: pertaining to the blood vessels.

Virus: even tinier than bacteria and they depend on a living host to reproduce. They cannot do it on their own due to their comparatively weak DNA.

W.

White Matter: so named because of the glistening myelin sheaths around or insulating the axons (of neurons) in the brain.

Working Memory: a type of memory where information is readily available for a very short period of time.

Y.

Yoga: there are many different kinds of yoga. But what sets these apart from other forms of exercise it that they aim to unite the mind and the body. The physical postures are intended not as the end goal, but as a way of occupying one's thoughts and liberating the mind.

Z.

Zone of Proximal Development: the space or zone between what a child is capable of doing on his/her own, and what he/she can achieve with assistance from an adult or more capable peer.

Index

action potential, 46, 131
adrenaline 35–36, 96
ageing 141, 175, 182
Alzheimer's disease 60, 142, 192, 208
amygdala 87, 97, 101, 213
antioxidants 173–174
attention 28–29, 74, 88–89, 152, 177
axon 44–45, 51, 131, 184

behaviourism 87
blood-brain barrier 174, 175
bloodstream 35, 169, 243
Bloom, Benjamin 74
blooms taxonomy 74–75, 132
brain health 79, 148, 173, 256
brain plasticity 50–51, 56–57, 66, 86
brainstem 30–31, 183
Bruner, Jerome 75, 77

Cajal, Ramon y 16, 56
case studies 59, 60, 77, 137, 187
central nervous system 41, 132, 168
cerebellum 24, 29–30, 133, 183
cerebral cortex 24–25, 27
cerebrospinal fluid 13, 187, 208
chunking 118–119
circadian rhythm 205, 206, 207, 208
cognition 20, 61, 74
cognitive development 73, 154
cognitive load 86, 88, 118,
cognitive psychology 75, 77, 99
cognitive reserve 61, 142–143
consolidation 87, 95, 97, 99–100, 198
corpus callosum 26

cortex 24–25, 27, 32–33, 87–88, 97, 131–132

demyelinated axons 131
dendrite 44–45, 58, 63, 133
Dewey, John 72–73
discovery learning 76
dopamine 97, 121, 135, 150, 214–215, 245
du Bois-Reymond, Emil 16
dual coding 87

empathy 152–153, 255–256, 259
endorphins 150, 191
enriched environment 58–59, 77, 185
emotions 32–33, 36–37, 97, 136, 151
executive function 97, 115, 161, 177
exercise 181–182, 186, 246

fatigue 177, 241
foetal development 132
food 165–166, 169, 171, 173
forebrain 24, 243
frontal lobe 19–20, 27, 137, 213, 216
fruit 173–174, 249
FMRI 76
Flavell, John 104

Gage, Phineas 19–20
Galen 15
Gall, Franz Joseph 17
Genes 44, 65, 134, 153, 171
Gibbs, G. 124
glial cells 51, 133, 176, 187, 273
glucose 35–36, 89, 165–166, 177
gut 167–168, 171–172

GYRI 28

habits 32, 64–65, 113–114, 123
habit formation loop 119–120
happiness 235–236
hearing 29, 90
Hebb, Donald, O. 50, 57–58, 85
hemispheres 24, 26–27
hippocampus 36, 60, 87, 101, 185
Hippocrates 14–15
Habituation 116
homeostasis 171
hormones 35, 97, 150, 205
hypothalamus 34–35, 150, 243
hydration 241, 243–244, 248

infant attachment 134, 139
intelligence 67, 138, 219, 263
interpersonal relationships 147–148, 151, 153
interleaved learning 92, 94, 95
intrinsic motivation 234–235
information processing theory 87

Kahneman, Daniel 88, 115
Kaizen principle 214, 218
Kuhn, Thomas 55

learning zone 64
long-term memory 89–90, 101, 234, 246
learning 29, 51, 71–73, 85–86, 92, 99
limbic system 32–33, 117, 135–136
London taxi drivers 59–60
loneliness 159–160

massed practice 93
Maslow, Abraham 232–233
melatonin 206, 210
memory 28–29, 37, 63, 85, 90, 100
metacognition 3, 104–105
midbrain 24

mild cognitive impairment 242
mindset 219, 221, 231
mirror neurons 151, 255, 258
motivation 37, 78, 98, 105, 115, 155, 229
    extrinsic motivation 234
movement 25–26, 29–30, 182, 183
magnetic resonance imaging (MRI) 1, 28, 56, 76
myelin 43, 64
myelin sheath 43, 45, 51, 130
myelination 45, 51, 130–131, 136

nervous system 31, 41–42, 118, 153, 168
neural network 43, 48–49, 65, 132
neural pathways 2, 36, 47, 56, 62
neural tube 132–133
neurobiology 71
neurogenesis 130, 181, 184–185, 186
neuroimaging 87, 89, 107
neuromyths 12, 78, 82
neuron 41–42, 43–44, 46
neuroplasticity 56–57, 62, 66, 214, 263
neuroscience 2, 17, 19, 56
neurotransmitters 46–47, 50–51, 149, 176, 209, 245
noradrenaline 96
nuclei 33
nucleus 88, 135, 219

occipital lobe 27, 29
omega-3 fatty acids 35, 176
oxytocin 150, 261

partnerships 155, 157
Parkinson's disease 32, 49, 208
patterning 99
performance 169, 177, 198, 220, 241, 245
phrenology 17, 18
physical movement 25–26, 29–30, 182, 183
Piaget, Jean 73–74, 75

Index

plasticity 50–51, 55, 57, 62, 131
pons 31
practice in learning 50, 65, 76, 93
prefrontal cortex 97, 117, 131, 135–136, 216
problem-based learning 235
proteins 47, 175–176, 201
pruning, synaptic 47, 58, 62, 64, 130–131

rapid eye movement (REM) 202–203
rehearsal 99–100, 101
reinforcement 50, 72, 92, 214
relationships 139, 147–148, 151–152, 155
resilience 157, 228–229, 232
reticular activation system 213
retrieval practice 93, 102
reward mechanisms 116–117, 120–121, 215, 222
Rosenzweig, Mark 58

salutogenesis 229
scaffolding 92
self-regulation 28, 97, 104, 256
sensory information 25, 29, 134, 183
sequencing 30, 52, 93–94
serotonin 176, 191, 245
short-term memory 63, 99–100, 118, 245
Sherrington, Charles 56
Skinner, B.F. 72, 75, 235
sleep 161, 197–198, 200–201, 202–203
smart targets 213, 217, 222
social interaction 74, 152, 235
socio-cultural learning 82
social brain hypothesis 148–149
social isolation 159–160
spaced practice 92
spinal cord 13, 31, 41

stress 35, 97, 115, 156, 160, 169, 228
stress response 35, 150
structure of the brain 23–24, 25–26
sulci 28
synapses 46–47, 50, 57–58, 76–77, 133–134
synaptic plasticity 50, 62
system thinking 115

targets 213–214, 220–221
temperature, body 150, 167, 203, 205
temporal lobe 27, 29, 38, 87
thalamus 32–33, 88, 98
theory of mind (TOM) 260
Thorndike, Edward 72
thinking 74, 104, 115, 119, 138
timing 30, 170,
transfer of learning 99, 103–104, 262

vagus nerve 168–169
van Leeuwenhoek, Anton 16
vascular 187
ventricles 15, 208
vitamins 167, 171, 241–242
Vygotsky, Lev 74–75, 135, 154

water warriors 248
Willis, Thomas 15
working memory 28, 85, 89–90, 209–210
writing 50, 105, 115, 119

youth risk-behaviour 89, 135, 137
yearning 227–228, 232, 235

zone of proximal development 74